New Directions in
DYNAMICAL SYSTEMS, AUTOMATIC CONTROL AND SINGULAR PERTURBATIONS

JOHN O'REILLY

Emeritus Professor of Engineering,
University of Glasgow, UK.

Matador
Unit E2 Airfield Business Park,
Harrison Road, Market Harborough,
Leicestershire. LE16 7UL
Tel: 0116 2792299
Email: books@troubador.co.uk
Web: www.troubador.co.uk/matador
Twitter: @matadorbooks

ISBN 978 1803132 013

British Library Cataloguing in Publication Data.
A catalogue record for this book is available from the British Library.

Printed and bound in the UK by TJ Books Limited, Padstow, Cornwall
Typeset in 11pt Minion Pro by Troubador Publishing Ltd, Leicester, UK

Matador is an imprint of Troubador Publishing Ltd

New Directions in
DYNAMICAL SYSTEMS,
AUTOMATIC CONTROL AND
SINGULAR PERTURBATIONS

Acknowledgement: The assistance of Hannah Dakin at Matador in the preparation of this book for publication is gratefully acknowledged.

CONTENTS

PART 1
DYNAMICAL SYSTEMS AND AUTOMATIC CONTROL

PART I

DYNAMICAL SYSTEMS AND AUTOMATIC CONTROL

PREFACE TO PART I

Part 1 of this book sets out the fundamental conditions that small-signal physically realisable dynamical system models must satisfy. These fundamental conditions are causality and non-singularity. They apply to all small-signal dynamical system models; for example, those that arise in electrical networks.

Another important example is automatic control. The key developments in twentieth-century automatic control are the works of Nyquist (1932) and Bode (1940) for systems with a single input and a single output. The paper by Nyquist (1932) in particular specifies causality, but otherwise confines its attention to the stability issue for which it is justly famous. Part 1 of this book, with its physical narrative and due care of historical context, reinterprets these classic works to establish that the uncontrolled system must also not be singular, nor must the controlled system encounter singularity.

However, Part 1 of this book goes much further. It shows that these fundamental properties – in particular, non-singularity – must obtain for all small-signal system models, regardless of how many inputs and outputs the system happens to have. Feedback must preserve the independence of the system outputs. So, small-signal automatic control is all of a piece. It is that simple. Moreover, it naturally leads to the coordinated decentralised control of large systems – for example, power systems, with their many inputs and many outputs – with consummate ease. The original methods of Nyquist (1932) and Bode (1940) then apply directly to all such multi-input multi-output control systems, without adaptation.

ONE

PHYSICAL DYNAMICAL SYSTEMS

SUMMARY
This chapter sets out the two fundamental properties that a physical dynamical system should possess. These two fundamental properties are, firstly, that the dynamical system should be physically realisable or causal, and secondly, that the outputs of the dynamical system should be independent.

1.1   INTRODUCTION

In every theoretical investigation of a real physical system, we are always forced to simplify and idealise, to a greater or smaller extent, the true properties of the system.

Theory of Oscillators
A. Andronov, A. A. Vitt and S. E. Khaikin (2011, P. xv)

In our quest to develop mathematical models for our chosen purpose, we often forget that these very models are but idealisations of the actual system we seek to describe. Such is the sentiment expressed previously in the opening sentence of Andronov, Vitt and Khaikin (2011) for physical dynamical systems in 1937 (in Russian).

Furthermore, this idealisation of physical dynamical systems must necessarily extend to their properties. It is with these limitations of model description in mind that we proceed to examine two fundamental properties of dynamical system models that, in their way, are most often merely assumed. These two fundamental properties are causality and independence of the outputs of the dynamical system under consideration. Our examination of these two properties has engineering application in its sights, but may also be of interest to the physical and biological sciences.

While the concept of a dynamical system took shape in Newtonian mechanics, it is Poincaré (1892–1899), in his study of celestial mechanics, whom is widely regarded as the pioneer of modern dynamical systems. Important work on the stability of dynamical systems was also presented in 1899 by Lyapunov (1992). The methods of Poincaré and Lyapunov remain the bedrock of contemporary studies of dynamical systems. Mention should also be made of the classic paper of Van der Pol (1926) on nonlinear oscillators as they arose in triode valves (tubes). Other important developments from both a mathematical and physical standpoint include the works of Birkhoff (1927) and Smale (1967).

But it is to publications with an engineering bent that our attention is now directed. The book of Andronov, Vitt and Khaikin (2011) on oscillators already referred to contains examples drawn from electrical circuits, pendulums and steam engines. Closer to the present day, the many publications of Nayfeh, among them Nayfeh and Mook (1979), on nonlinear dynamical systems have had an enormous inpact on structural mechanics as they arise in aircraft, ships and engineering structures of all kinds. Likewise, the work of Chua (1969) on nonlinear network theory has exerted a great influence on electrical engineering. For a recent introduction to nonlinear networks, see Muthuswamy and Banerjee (2019). Structural mechanics and nonlinear networks may be at opposite ends of the engineering spectrum, but they often share the common property of possessing a physical dynamical system model.

Chapter 1 is organised as follows. Section 1.2 defines what is meant by a dynamical system model, in both its nonlinear and linearised form, with an example drawn from robotics. The fundamental physical system properties of causality and independence of system outputs are then presented in Section 1.3; a most useful application to automatic control of these causality and independence properties is provided in Chapter 2. Conclusions are outlined in Section 1.4.

1.2   DYNAMICAL SYSTEM MODELS

Dynamical models of physical systems are invariably nonlinear in their dynamical behaviour. They can be represented by the following system of ordinary differential equations, with initial condition $x(t_0) = x^0$, given by

$$\frac{dx}{dt} = f(x,u), \quad x(t_0) = x^0, \quad x(t) \in R^n, \quad u(t) \in R^m \tag{1.1}$$

$$y = h(x), \quad y(t) \in R^m \tag{1.2}$$

where $t \geq t_0$ denotes time, $x(t)$ is the system state vector, $u(t)$ is a vector of system inputs and $y(t)$ is the corresponding vector of system outputs.

It is observed in the dynamical system model (1.1) and (1.2) that there is an equal number m of system inputs and system outputs. Each of these m system outputs is paired with a particular system input. The most appropriate pairing of system input to system output is determined by the physical context.

For the nonlinear dynamical system model (1.1) and (1.2) to be physically realisable, it is required that the system model be causal; in other words, the system output $y(t)$ can only depend upon past and present system inputs $u(\tau)$, $\tau \leq t$, but not upon future system inputs $u(\tau)$, $\tau > t$.

Also, as noted, the nonlinear dynamical system model (1.1) and (1.2) contains several inputs (forcing functions) and several outputs (responses). It is therefore also reasonable to require that the system outputs can be manipulated independently by suitable choices of system inputs; for example, as in a robotic arm (Zivanovic and Vukobratovic, 2006). There seems little point in considering any system output that depends upon other system outputs. The following example illustrates these points.

Example 2.1: Consider the two-link planar manipulator robot arm depicted in Figure 1.1. The equations of motion that describe the dynamical response of the manipulator arm to input joint torques take the form of the nonlinear dynamical system model (1.1) and (1.2). The two system inputs are the joint actuator torques, τ_1 and τ_2, provided by electrical motors and applied to Joint 1 and Joint 2, respectively. The two corresponding system outputs – the motions, so to speak – are the link angular rotations, θ_1 and θ_2, in each of the two links of the manipulator.

Naturally, the nonlinear manipulator arm system of Figure 1.1 is physically realisable or causal. There can be no change in the link angular rotations θ_1 and θ_2 until after a change in the actuator torque inputs τ_1 and τ_2 has occurred. No output response occurs before an input stimulus is applied.

Moreover, it is entirely reasonable to wish to manipulate independently the output angular rotations θ_1 and θ_2 in each of the two links by way of appropriate joint actuator torques τ_1 and τ_2.

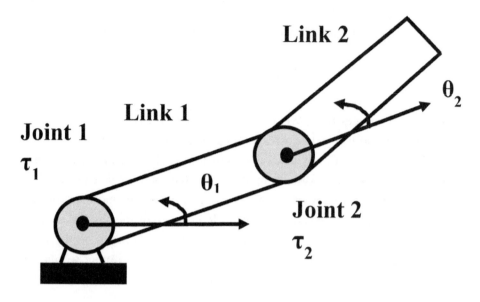

Figure 1.1 A two-link planar manipulator robot arm

In this short chapter on physically realisable dynamical systems, these ideas assume a more concrete form if we focus the discussion on the small-signal or linearised system models resulting from the linearisation of (1.1) and (1.2) about a suitable operating point.

Let x_e, u_e and y_e be the equilibrium values of x, u and y, respectively, of the nonlinear dynamical system (1.1) and (1.2) for a constant input u_e; that is,

$$f\left(x_e, u_e\right) = 0 \tag{1.3}$$

$$y_e = h\left(x_e\right) \tag{1.4}$$

Consider small deviations

$$\delta x = x - x_e, \quad \delta u = u - u_e, \quad \delta y = y - y_e \tag{1.5}$$

about the equilibrium values $\left(x_e, u_e, y_e\right)$. By Taylor series expansions, where only first-order terms are retained, these small deviations δx, δu and δy about the equilibrium values $\left(x_e, u_e, y_e\right)$ are approximately described by the small-signal or linearised dynamical model

$$\frac{d(\delta x)}{dt} = \frac{\partial f}{\partial x}\delta x + \frac{\partial f}{\partial u}\delta u, \quad \delta x(t_0) = x^0 - x_e \quad (1.6)$$

$$\delta y = \frac{\partial h}{\partial x}\delta x \quad (1.7)$$

where the partial derivatives $\partial f / \partial x$, $\partial f / \partial u$ and $\partial h / \partial x$ – evaluated at $x = x_e, u = u_e$ and $y = y_e$ – are constant matrices. Properties that are local to the equilibrium point (x_e, u_e, y_e) of the original nonlinear dynamical system (1.1) and (1.2) can be deduced from the linearised dynamical model (1.6) and (1.7). It is therefore to the small-signal model (1.6) and (1.7) that we henceforth confine our attention.

It is convenient to rewrite the small-signal or linearised dynamical model (1.6) and (1.7) as the linear time-invariant dynamical system model

$$\frac{dx}{dt} = Ax + Bu, \quad x(t_0) = x^0, \quad x(t) \in R^n, \quad u(t) \in R^m \quad (1.8)$$

$$y = Cx, \quad y(t) \in R^m \quad (1.9)$$

which differs from (1.6) and (1.7) only in the obvious change of notation with constant matrices A, B and C.

It is also the case, as we shall see in Section 1.3, that the linear time-invariant dynamical system model (1.8) and (1.9) exhibits further interesting properties when transformed into the complex domain. Taking one-sided Laplace transforms of (1.8) and (1.9), we have

$$sx - x^0 = Ax(s) + Bu(s) \quad (1.10)$$

$$y(s) = Cx(s) \quad (1.11)$$

in which $s = j\omega$ denotes the complex variable. Ignoring the initial conditions, it is then a simple matter to rearrange the model (1.10) and (1.11) as the strictly proper transfer-function matrix $G(s)$ from the input $u(s)$ to the output $y(s)$ given by

$$y(s) = G(s)u(s) \quad (1.12)$$

9

where

$$G(s) = C(sI - A)^{-1} B \tag{1.13}$$

Of course, the transfer-function elements of $g_{ij}(s)$ of the m-input m-output transfer-function system matrix $G(s)$ of (1.12), as depicted in Figure 1.2, may either come from the differential equation system model (1.10) and (1.11), if known, or experimentally from measured frequency-response data.

Since, as noted previously, it makes no sense for the system model $G(s)$ of (1.12) to possess either redundant system inputs or redundant system outputs, the system model matrix $G(s)$ is a square m x m matrix.

Henceforth, we consider in the complex domain lumped m-input m-output linear time-invariant dynamical systems of the form described by the square m x m transfer-function matrix $G(s)$ of Figure 1.2 where s is the complex variable $s = j\omega$. It is assumed in Figure 1.2 that $G(s)$ is rational in s over the field of complex numbers; that is to say, each transfer-function element $g_{ij}(s)$ of $G(s)$ is a rational function of s possessing numerator and denominator polynomials.

I.3   LINEAR PHYSICAL DYNAMICAL SYSTEMS

In this section, we shall be working with the lumped transfer-function system depicted in Figure 1.2 which may arise from either a linearised differential equation system model (1.10) and (1.11), or from measured frequency-response data. First of all, it is necessary to define what is meant by a 'linear physical dynamical system'. Here, we are using the terms 'physical system' and the 'linear dynamical model' for such a system interchangeably.

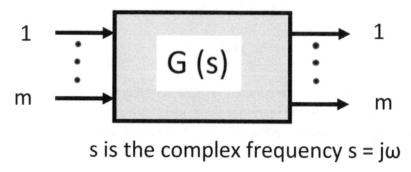

Figure 1.2 A lumped m-input m-output transfer-function system

Definition 3.1: The necessary and sufficient conditions for a linear physical dynamical system to exist are as follows:

i. For a given set of linearly independent system inputs, the set of system outputs is linearly independent.
ii. The linear dynamical system is physically realisable.

Geometrically, Condition (i) and Condition (ii) of Definition 3.1 for a physically realisable linear dynamical system can be visualised as in Figure 1.3.

Definition 3.1 is fundamental to *all* linear physical dynamical systems, including those drawn from dynamics, automatic control and network theory. Condition (ii), of course, is well known (Nyquist, 1932; Kalman, 1963) and requires that the dynamical system be causal or non-anticipative, in the sense that the system outputs can only depend upon past and current system inputs, but *not* on future system inputs. It therefore excludes non-causal systems, which are anticipative, such as those that are to be found in some signal-processing algorithms.

Condition (ii) of Definition 3.1 is rooted in the laws of classical mechanics and electromagnetism that are associated with Newton, Maxwell and others. Put simply, for a given input stimulus to the dynamical system, the system delivers a corresponding output response. Should there be more than one input to the linear dynamical system, Condition (i) of Definition 3.1 means that to each and every system input stimulus, there is a corresponding system output response. Put another way, Condition (i) specifies that there

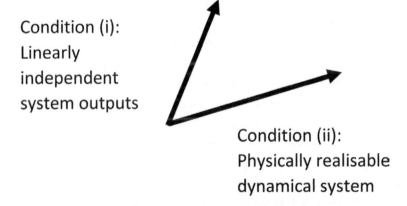

Condition (i):
Linearly
independent
system outputs

Condition (ii):
Physically realisable
dynamical system

Figure 1.3 A linear physically realisable dynamical system

are no redundant or dependent system outputs. Dynamicists will certainly be familiar with Condition (ii). Control engineers assume Condition (ii), with occasional reference to Condition (i). For instance, Condition (i) is assumed in Rosenbrock (1969), O'Reilly (1983), and Fahmy and O'Reilly (1988).

A physicist or mathematician may nod their head at Condition (ii), but wonder as to why Condition (i) was included in Definition 3.1. The engineer, on the other hand, is more likely to be alert to the possible engineering objectives served by Condition (i) in the linear dynamical system possessing independent outputs for a given set of linearly independent system inputs. For instance, the dynamicist may well be interested in manipulating independently certain angular motions of the links in a robotic arm, as in Figure 1.1 of Example 2.1.

Of course, the prior assignment of particular system inputs to particular system outputs in Condition (i) of Definition 3.1 constitutes an ordered integer numbering of system inputs and outputs. With that in mind, let us formalise the ordering of the system inputs and system outputs in Condition (i) as permutations among an ordered set of integer numbers in the manner of the opening definition of Mirsky (1955, P. 2).

Definition 3.2: Suppose the ordered sets $(\lambda_1, \ldots, \lambda_m)$ and $(\mu_1, \ldots, \mu_m)$ contain the same distinct integers, but these integers do not necessarily occur in the same order, then $(\lambda_1, \ldots, \lambda_m)$ and $(\mu_1, \ldots, \mu_m)$ are said to be permutations of each other. In symbols, $(\lambda_1, \ldots, \lambda_m) = P(\mu_1, \ldots, \mu_m)$ or $(\mu_1, \ldots, \mu_m) = P(\lambda_1, \ldots, \lambda_m)$.

In other words, if for any reason a particular assignment of system inputs to system outputs proves to be unsatisfactory, the system inputs and outputs can be reassigned at will, without loss of generality. Most often, the appropriate choice of input-output assignment is dictated by long engineering experience – indeed, engineering practice – in a particular industry.

In Example 2.1, on the robotic manipulator arm, it makes physical sense to pair the input torque τ_1 with the output angular rotation θ_1 in Link 1; likewise, to pair the input torque τ_2 with the output angular rotation θ_2 in Link 2.

As a further example from electrical power systems, consider the turbogenerator connected to an infinite bus, as in Fadlalmoula et al. (1998). The turbogenerator is a two-input two-output electromechanical dynamical system where the two outputs are the terminal voltage and shaft speed. The two inputs are provided by the excitation-automatic voltage regulator and the turbine governor. More than a century of power system experience suggests, unsurprisingly, that the electrical input and output should be paired, and that the mechanical input and output should be paired. The alternative pairing was investigated theoretically by Fadlalmoula et al. (1998), but it was discounted because of the practical disadvantages of a lack of system integrity and the subjection of the governor to high-frequency signals.

Other personal experience with helicopters suggests that getting the initial assignment of system inputs to system outputs wrong can turn a straightforward engineering dynamical problem into a much more difficult or impractical one. In all cases within the author's experience, retaining a physical sense of what the system inputs and system outputs are avoids all such pitfalls. Further discussion, however, of the assignment of input-output pairings – in automatic control, in this instance – is to be found in Leithead and O'Reilly (1992a).

Assumption 3.1: The linear physical dynamical system of Definition 3.1 is unconditionally stable.

Assumption 3.1 means that, should the dynamical system be perturbed by an external input (intended or a disturbance), it will subsequently return to its equilibrium state. Dynamicists or network specialists will barely pause to peruse Assumption 3.1, since it accords with their engineering experience. Industrial control engineers too usually find (Rosenbrock, 1969) that their dynamical systems, upon examination, are naturally stable in their uncontrolled or open-loop system state (one notable exception is highly manoeuvrable military aircraft).

Basic physics reveals why this should be so. Physical systems are dissipative. That is to say, they dissipate energy to their surroundings. In particular, mechanical and electrical waves and oscillations lose energy over time, typically from such effects as friction, resistance and turbulence. The dynamical system model may capture some of these dissipative effects; more likely, the actual energy dissipation will be greater than that predicted by the model, due to the inevitable unmodelled nonlinear and distributed dynamical effects.

Let us now apply Definition 3.1 to the lumped m-input m-output linear time-invariant dynamical systems of the form described by the square $m \times m$ transfer-function matrix $G(s)$ of Figure 1.2 or equation (1.12), where s is the complex variable $s = j\omega$.

Result 3.1: The necessary and sufficient conditions for the lumped linear time-invariant physical dynamical system described by the m-input m-output square transfer-function matrix $G(s)$ in Figure 1.2 to exist are as follows:

i.   The $m \times m$ transfer-function matrix $G(s)$ is non-singular.
ii. The $m \times m$ transfer-function matrix $G(s)$ is strictly proper.

Remark 3.1: Condition (i) of Result 3.1, in that the square system transfer-function matrix $G(s)$ be non-singular, is equivalent to the condition in Definition 3.1 that, given the m system inputs are linearly independent, the m system outputs are also linearly independent. (The square $m \times m$ transfer-function matrix $G(s)$ is of full rank and hence non-singular.)

Remark 3.2: In Condition (ii) of Result 3.1, physical realisability for lumped linear time-invariant systems is equivalent to the condition that the transfer-function matrix $G(s)$ be strictly proper. This means that each transfer-function element $g_{ij}(s)$ of $G(s)$ is strictly proper in the sense that the degree of the denominator polynomial is greater than the degree of the numerator polynomial. For example, the state-space model (1.13) is strictly proper.

Remark 3.3: So as to avoid an ill-defined system (and an ill-defined linear independence property) in Result 3.1, it is also necessary that the square transfer-function matrix $G(s)$ be not nearly singular either. Put another way, the polar plot of $Det\ G(s)$ should not be close to the origin of the complex plane.

1.4   CONCLUSIONS

This short introductory chapter has laid out the two fundamental properties that a physical dynamical system should possess. These two fundamental properties are, firstly, that the dynamical system should be physically realisable or causal, and secondly, that the outputs of the dynamical system should be independent.

While these fundamental properties are assumed in many areas of dynamical systems, such as robotics or electrical networks, they have particular relevance for controlled dynamical systems, as described in Chapter 2. This is scarcely surprising since the roots of automatic control in the modern industrial era – associated with the stability results of Routh, Hurwitz, Maxwell, Lyapunov and others – lie in nineteenth-century dynamical systems (Mayr, 1971a, 1971b; MacFarlane, 1979). If Watt, with his earlier flyball speed governor for steam engines, was the inventor of automatic control in the modern era, Maxwell (1868) was its founder as a discipline (Mayr, 1971b). How this chapter's fundamental dynamical system properties of causality and independence of system outputs are intimately connected to automatic control is revealed fully in Chapter 2. Put simply, automatic control will be shown to be an adjunct of the physical dynamical systems described in this chapter.

TWO

TOWARDS A MORE PHYSICAL AUTOMATIC CONTROL

SUMMARY

This chapter is an affirmation of the greatness of automatic control. It advocates a new physically based approach towards it – a new direction. It is argued that automatic control is an engineering discipline whereby engineering objectives are to be met. Based upon first principles that the system outputs be independent and that the linear dynamical system be physically realisable (as established in Chapter 1), several notable control problems are reinterpreted. The key underlying issue of automatic control is identified as one of singularity and how to avoid it. This physically based approach also results in a new coordinated decentralised control framework, thereby directly addressing the complexities of modern industrial control practice. Another significant benefit of this more physical approach is that it lends itself to the solution of new technical problems of immediate interest to control theorists and engineers alike.

2.1   INTRODUCTION

Since the days of antiquity, it has been the privilege of the mathematician to engrave his conclusions, expressed in a rarefied and esoteric language, upon the rocks of eternity. While this method is excellent for the codification of mathematical results, it is not so acceptable to the many addicts of mathematics, for whom the science of mathematics is not a logical game, but the language in which the physical universe speaks to us, and whose mastery is inevitable for the comprehension of natural phenomena.

Linear Differential Operators
C. Lanczos (1996, P. xiii)

This memorable quotation from the Preface of Lanczos (1996), originally published in 1961, serves as a reminder that new results in any field need to be communicated effectively if they are to prove useful.

This chapter, with its new results, is first and foremost an engineering chapter. With its stress on the fundamentals of physical dynamical systems established in Chapter 1, its originality lies chiefly in the advocacy of a more physical approach to automatic control. It thereby establishes a fundamentally new direction for automatic control. The germ of this idea for such a new physical direction came early in the author's career when, as a young man, he stumbled upon the book *Personal Knowledge* by the philosopher Michael Polanyi (1973). There, it is cogently argued that, in contradistinction to the sciences, such as physics and chemistry, engineering must perforce obey higher organising principles. So, while a wheelbarrow can be said to obey Newton's laws in its motion, the same laws can never, on their own, invent a wheelbarrow. The invention itself can only come about when the engineering artefact is designed, harnessing physical laws, to meet some other specified need or purpose.

A striking example of this other specified need or purpose, from almost a century ago, that was later to have a profound effect on the field of automatic control is Harold Black's invention of the negative-feedback amplifier (Black, 1934). In his own recollection, Black (1977, P. 59) states, "I suddenly realised that if I fed the amplifier output back to the input, in reverse phase, and kept the device from oscillating (singing as we called it then), *I would have exactly what I wanted* [italicisation added for emphasis]: a means of cancelling out the distortion in the output." Black had a purpose. So too does automatic control.

The purpose of automatic control is to meet control objectives. Naturally, these control objectives will vary from industry to industry. Generally speaking, typical control objectives might include zero system steady-state error, satisfactory system transient response, insensitivity to model uncertainties and external disturbances, and insensitivity to noise and other unmodelled high-frequency dynamics.

In many ways, automatic control is a hidden technology. It does its important job unobserved. One only notices it when it fails, as in an aircraft crash or in a large-area electrical power system outage. However, it was not always so; witness the early days of controlled flight and ship steering control. These early public successes demonstrate that it is helpful if one identifies the control device and names it, as with Black's negative-feedback amplifier. A physical appreciation as to the purpose – what it is for – and the principle of operation of the automatic control device is essential.

In this way, it is observed that control engineering objectives and dynamical system properties are, first of all, qualitative in nature. These qualitative considerations relate to the interactive character of the design process itself. Once the control objectives and system properties are thereby suitably identified, considerations of a more quantitative nature come into play. These further quantitative considerations relate to design accuracy and within how much margin of error the design should meet with its stated objectives.

Taking a more physically based approach to dynamical systems, this chapter advances the argument that automatic control is an engineering discipline whereby engineering objectives are to be met. The emphasis is upon first principles. Specifically, as revealed in Chapter 1, it is required that the system outputs be independent, and that the linear dynamical system be physically realisable. The key underlying issue of automatic control is thereby identified as one of singularity and how to avoid it. This provides a powerful physical context for the reinterpretation of several notable control problems. Two such problems, deserving of being better known in automatic control, are the phase-locked loop (PLL) and the power system stabiliser (PSS).

The development is simple, direct and progressive. There are considerable benefits to the approach. One of them is that multi-input multi-output control problems are not so very different to single-input single-output (SISO) problems. It also becomes obvious that, in spirit, the approach lends itself to the solution of new technical problems that are of interest to control theorists and engineers alike. Throughout, as in Chapter 1, this chapter is guided by first principles. We have not turned our back on empiricism, though; engineering experience and industrial practice are also shown to have a valuable role to play. Together, first principles deduction and engineering empiricism form a most physical approach to automatic control. Indeed, we are mindful that, "However beautiful the strategy, you should occasionally look at the results."[1]

---

1 Though often attributed to Winston Churchill, it appears that this saying was first uttered by the UK Conservative politician Ian Gilmore in 1981.

Regarding the chapter itself, an acquaintance with elementary linear algebra is helpful, as is an appreciation of basic physics.

The chapter is organised as follows. Section 2.2 specialises the first principles results of Chapter 1 to negative-feedback automatic control. Starting with SISO systems, the succeeding Sections 2.3, 2.4 and 2.5 use the results of Section 2.2 progressively to reinterpret, in a more physical manner, a number of important control problems. Some are well known; others less so. Section 2.6 on coordinated decentralised control is new, as is Section 2.7 on performance specifications and control design. Concluding remarks are presented in Section 2.8. Indeed, the chapter can be read in a number of different ways. Those interested in the fundamentals of automatic control from a more physical viewpoint need only read Sections 2.2 and 2.3. For those interested in more historical material on ingenious control devices, Sections 2.3, 2.4 and 2.5 are recommended. The industrial engineer seeking guidance on the control of systems with more than one input and output need look no further than Sections 2.6 and 2.7. Others of a more leisurely bent should close their eyes momentarily, relax and enjoy the whole *singular* experience!

2.2 NEGATIVE-FEEDBACK AUTOMATIC CONTROL

It is at this point that we focus our discussions of physical dynamical systems as they arise in negative-feedback automatic control. Extensive use will be

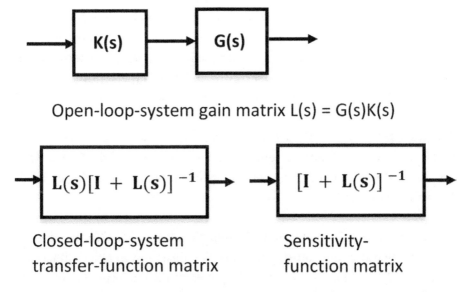

Open-loop-system gain matrix L(s) = G(s)K(s)

Closed-loop-system transfer-function matrix

Sensitivity-function matrix

Figure 2.1 Open-loop-system and closed-loop-system transfer-function matrix relations

made of the results of Chapter 1 on physical dynamical systems; in particular, Result 3.1.

Dispensing with the formalities, consider the familiar unity negative-feedback control configuration summarised in Figure 2.1. All matrices in Figure 2.1 are square $m \times m$ matrices. Were there to be a square $m \times m$ transducer matrix $H(s)$ in the feedback loop, it would be rolled into the $m \times m$ open-loop transfer-function matrix as $L(s) = G(s)H(s)K(s)$. The following result is immediate.

Result 2.1: Under the conditions of Result 3.1 of Chapter 1, namely the non-singularity of the square $m \times m$ plant matrix $G(s)$ and system causality, an m-input m-output unity negative-feedback controller with square $m \times m$ loop-gain matrix $L(s) = G(s)K(s)$ exists if and only if the square $m \times m$ return-difference matrix $[I + L(s)]$ of Figure 2.1 is non-singular.

This fundamental result is nothing more than a restatement in feedback form of the linear independence of system outputs and causality of Result 3.1 in Chapter 1. In other words, automatic control, with its m-input m-output feedback loops, is just a particular configuration of the physical dynamical system defined in Chapter 1. Feedback must therefore preserve the linear independence property of the m system outputs. Otherwise, a feedback control system ceases to exist as a physical system.

By Result 2.1, should it be the case that the square $m \times m$ return-difference matrix $[I + L(s)]$ is singular, both the closed-loop-system transfer-function matrix and the sensitivity-function matrix of Figure 2.1 are undefined. This naturally means that all properties of the closed-loop-system transfer-function matrix and sensitivity-function matrix are undefined. The control objectives are also undefined should the matrix $[I + L(s)]$ prove to be singular.

Remark 2.1: By Result 3.1 of Chapter 1, where the square $m \times m$ plant matrix $G(s)$ is non-singular, and the square $m \times m$ controller matrix $K(s)$ is non-singular by design, the square $m \times m$ loop-gain matrix $L(s) = G(s)K(s)$ in Result 2.1 is also non-singular.

Remark 2.2: Result 2.1 builds upon the non-singularity of $G(s)$ in Result 3.1 of Chapter 1 by way of the non-singularity of the square $m \times m$ return-difference matrix $[I + L(s)]$ of Figure 2.1. It thereby preserves the independence of inputs and independence of outputs round the negative-feedback loop.

Result 2.1, Remark 2.1 and Remark 2.2, on non-singular or well-defined feedback systems, have their natural corollary in ill-defined feedback systems, as described in the next remark.

Remark 2.3: So as to avoid an ill-defined feedback system (and ill-defined feedback properties) in Result 2.1, it is also necessary that the square $m \times m$ return-difference matrix $[I + L(s)]$ of Figure 2.1 be not nearly singular either. Put another way, the polar plot of $Det [I + L(s)]$ should not come close to the origin of the complex plane.

Under the conditions of Result 3.1 of Chapter 1 and Result 2.1, where the loop transfer-function matrix $L(s)$ is strictly proper, two properties of the closed-loop-system transfer-function matrix in Figure 2.1 are immediate.

Properties 2.1:

i.   For $L(s)$ sufficiently large (at low frequency), the $m \times m$ closed-loop-system transfer-function matrix $L(s)\left[I+L(s)\right]^{-1}$ tends to the identity matrix I_m.

ii. For $L(s)$ sufficiently small (at high frequency), the $m \times m$ closed-loop-system transfer-function matrix $L(s)\left[I+L(s)\right]^{-1}$ tends to the null matrix O.

2.3   SINGLE-LOOP CONTROL

The most common form of automatic control by far is single-loop control where there is just one system input and one system output to consider. With such a single-loop control, all the matrices in the system description of Figure 2.1 collapse to scalars; specifically, the plant transfer function $g(s)$ and the system open-loop transfer function $l(s)$ are now scalar rational functions of the complex variable $s = j\omega$.

As a consequence, Result 3.1 of Chapter 1, Remark 2.1 and Result 2.1, respectively, inform us that the transfer functions $g(s)$, $l(s)$ and $[1 + l(s)]$ should all be non-zero. In particular, keeping Assumption 3.1 of Chapter 1 in mind, the polar plot of the open-loop-system transfer function $l(s)$ typically starts at zero (angular) frequency on the positive real axis of the complex plane, as sketched in Figure 2.2. Since by Condition (ii) of Result 3.1 of Chapter 1, $l(s)$ is also strictly proper, then $l(s)$ must roll off to zero at high frequency; again, as sketched in Figure 2.2.

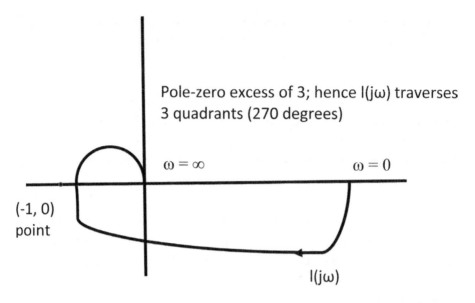

Pole-zero excess of 3; hence l(jω) traverses 3 quadrants (270 degrees)

ω = ∞

ω = 0

(-1, 0) point

l(jω)

Figure 2.2 Polar plot of a typical stable open-loop system frequency response l(jω)

It remains to be determined what happens to the polar plot of the open-loop-system transfer function $l(s)$ at frequencies in between low frequency and high frequency. That part of the polar plot will increasingly depend on the pole-zero excess of the strictly proper loop transfer function $l(s)$ as frequency increases. Suppose, for example, that the pole-zero excess of $l(s)$ is three. This means that, as frequency increases, the polar plot of $l(s)$ will quite dramatically rotate clockwise 270 degrees through three quadrants of the complex plane. This drama is solely down to the 270-degree change of phase in $l(s)$ as frequency increases. Since by Assumption 3.1 of Chapter 1, $l(s)$ is stable, then the 270-degree rotation is clockwise. We now have a complete polar plot of a typical loop transfer function $l(s)$ as frequency increases, as sketched in Figure 2.2.

Such a polar plot as in Figure 2.2 might well have been sketched as early as in the 1920s by a number of authors predating Black's feedback amplifier invention of 1927 (Macfarlane, 1979). What is more, many of these same authors will doubtless have identified $l(s) = -1$ as a singular point of the closed-loop-system transfer function $l/[1 + l]$ in accordance with Result 2.1.

The closed-loop-system transfer function $l/[1 + l]$ and all its properties are undefined at this point; indeed, the original control objectives are undefined at this point. In other words, the point (−1, 0) on the complex plane sticks out like a sore thumb; this point is singular and so needs to be avoided by the polar plot of $l(s)$ at all costs. As the loop gain $l(s)$ passes through critical point $(-1, 0)$, the scalar return-difference $1 + l(s)$ passes through zero, and hence experiences a change of sign or a 180-degree shift in phase. These comments on ill-defined systems are thrown into sharper relief in the following remark.

Remark 2.4: In accordance with Remark 3.3 of Chapter 1 and Remark 2.3, so as to avoid an ill-defined feedback system (and ill-defined feedback properties), it is also necessary that the plant transfer function $g(s)$ and the scalar return-difference $1 + l(s)$ be not near zero either. Put another way, the polar plot of $g(s)$ should not go near the origin of the complex plane, and the polar plot of $l(s)$ should not go near the $(-1, 0)$ point of the complex plane.

Looking at the polar plot sketched in Figure 2.2, the designer, on the face of it, appears to have done a good job in avoiding the singular point $(-1, 0)$ by having the plot of $l(s)$ pass it well to the right in its roll off to the origin. In many cases, the designer will indeed have done a good job. However, there is a sting in the tail if the pole-zero excess in $l(s)$ is considerable: the gain of $l(s)$ may roll off too quickly, such that the pole-zero excess may induce an unwelcome clockwise swing at the singular $(-1, 0)$ point. A dramatic phase change again, I am afraid. Hence Bode's admonishment that the loop gain cannot be reduced too abruptly (Bode, 1940). But maintaining an eye on the polar plot keeps the designer right! And the beauty of it all is that the gain and phase of the open-loop system $l(s)$ in its polar plot can be determined directly from frequency response measurements. It sounds almost too good to be true, so what's the catch?

The catch is that there is a cost attached to feedback. The primary cost of feedback, as Black's colleagues would have reminded him (Mindell, 2000), lies in the reduction of hard-won system gain. Specifically, Figure 2.1 shows us that the closed-loop-system gain is reduced by a factor of $1/[1 + l]$ as compared to the open-loop-system gain. But there are other costs to feedback that Bode (1940) cheerfully reminds us of at length, so that the feedback enthusiast may be deterred from rash commitments. These costs have already been hinted at in our discussion of control objectives and, indeed, in Black's (1977) recollection. Aside from amplifier gain reduction, Black's (1977) cost was amplifier oscillation instability.

Other costs, mentioned herein in the control objectives, are sensitivity to noise and unmodelled high-frequency dynamics. Increased system bandwidth might also be considered a cost. On the other hand, the primary benefits of feedback control, among control objectives, are zero system steady-state error, satisfactory system transient response, and insensitivity to model uncertainties and external disturbances. In short, the benefits of feedback control accrue from high open-loop system gain, but at the cost of increased instability, increased sensitivity to noise, and unmodelled high-frequency dynamics.

This brings us to one of those costs of feedback: instability of the resulting closed-loop system. The good news, of course, is Assumption 3.1 in Chapter 1. Almost all systems start off stable before the feedback specialist arrives. It is reasonable to suppose that given a system that is stable to start with, any open-loop-system $l(s)$ polar plot that avoids the singular point $(-1, 0)$ in the manner described previously should have no problem with closed-loop-system instability. This supposition would have been confirmed by experiment. In fact, despite Black's recollection (Black, 1977) to the contrary, Mindell (2000), in his thorough historical analysis, is at pains to stress that in Black's 1928 and 1932 patent applications there is little or no mention of the possibility of amplifier oscillation instability.

It is here that Nyquist entered the picture when he asked Black in 1928 to join in the development of a new telephony carrier system that included Black's newly invented feedback amplifier (Mindell, 2000). Nyquist was profoundly interested in the issue of possible closed-loop-system instability of the amplifier and gave, with proof, the rule to avoid it (Nyquist, 1932); here is the rule *verbatim*.

Rule: Plot plus and minus the imaginary part of AJ(iω) against the real part for all frequencies from 0 to ∞. If the point 1 + i0 lies completely outside this curve the system is stable; if not it is unstable. (Nyquist, 1932, P. 136)

Allowing for a sign change due to the feedback amplifier configuration, the Nyquist rule in our notation is that the polar plot of $l(j\omega)$ should not enclose the $(-1, 0)$ point for stability. So, in our physically based loop-shaping discussion, which is actually closer to the spirit of Bode (1940) and starting from natural physically based assumptions and deductions, we almost got there without Nyquist. But not quite! It is the case that, of all the feedback properties mentioned, closed-loop-system transient response and stability assume the greater importance in the mid-frequency range where $l(s)$ may be close to the critical $(-1, 0)$ point. Enhanced closed-loop-system transient response demands higher open-loop-system gain $l(s)$, but enhanced closed-loop-system stability requires the opposite. Nyquist's stability rule, quite simply, is that if the designer wishes to retain system stability, they must not let the polar plot of $l(j\omega)$ go near (enclose) the $(-1, 0)$ singular point. It can happen, of course, that the plant does not satisfy Assumption 3.1 of Chapter 1 and is unstable. The plant may even be non-minimum-phase. These more difficult issues are explored thoroughly in Leithead and O'Reilly (1991).

Bode's supreme achievement was to integrate Black's negative-feedback-amplifier design objective with Nyquist's stability rule, so as to forge a coherent feedback design methodology. This feedback design methodology is what came to be known as 'loop shaping' and remains to this day the basis of automatic control.

2.4   PHASE-LOCKED LOOP (PLL)

The previous section, in pursuing a more physical interpretation of automatic control, highlighted the enormous achievements of Black, Nyquist and Bode. None of this trio, however, were control engineers; there is no reference whatsoever to automatic control in their original publications or patents, only to amplifiers. Why should there have been? Black, Nyquist and Bode were young telephone engineers at the forefront of bringing long-distance telephony to an eager public. Moreover, they were working alongside hundreds of other engineers at the premier company Bell Telephone Laboratories (AT&T) in New York, which was then, in 1927, the cultural and financial epicentre of the leading world industrial power undergoing a stock market boom. It must have been exciting. It was incidental that later Second World War fire-control demands subsequently assured the rapid assimilation of these negative-feedback amplifier concepts into automatic control (Macfarlane, 1979).

Nevertheless, we turn now to a control system invented two years earlier in 1925, which is no less well known to electronics and communications engineers, but is strangely neglected in the automatic control field; however, see Clarke and Park (2003). I refer to the ubiquitous phase-locked loop (PLL) originally invented by David Robertson, a professor at Bristol University UK, so as to synchronise the university pendulum clock with an external time signal telegraphed daily from Greenwich, London. As to what the PLL does, the clue lies in the name. Like the negative-feedback amplifier of Section 2.3, this device satisfies Result 3.1 of Chapter 1, Result 2.1 and Assumption 3.1 of Chapter 1.

The basic purpose of a PLL is to synchronise or control the phase (frequency) of a signal to the phase of a reference signal. In its simplest form, the PLL assumes the negative-feedback arrangement of Figure 2.3. The oscillator generates an

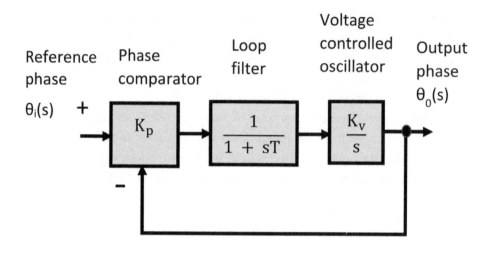

Loop gain $l(s)$

$$= \frac{K}{s(1 + sT)} \qquad ; K = K_p K_v \text{ and } T = RC$$

Figure 2.3 The phase-locked loop (PLL)

output signal whose phase is compared with the reference phase of an input signal by way of the phase comparator. The loop filter controls the oscillator so as to regulate the output signal phase to the reference input signal phase.

From Figure 2.3, the transfer function of the closed-loop system takes the form of the second-order stable system described by

$$l/(1+l) = \frac{\omega_n^2}{\left(s^2 + 2\xi\omega_n s + \omega_n^2\right)} \tag{2.1}$$

with natural frequency ω_n and damping ratio ξ, respectively given by

$$\omega_n = \sqrt{K/T} \quad , \quad \xi = 1/2\sqrt{KT} \tag{2.2}$$

It is observed that, under negative feedback, the resulting closed-loop system in (2.1) behaves as a damped oscillator.

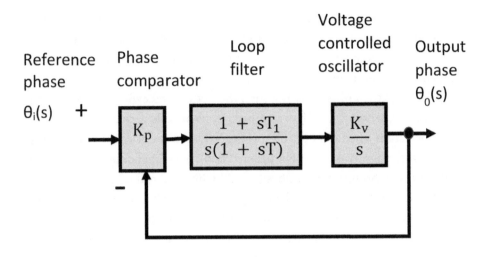

Loop gain $l(s)$

$$= \frac{K(1 + sT_1)}{s^2(1 + sT)} \qquad ; K = K_pK_v \text{ and } T = RC$$

Figure 2.4 The phase-locked loop (PLL) for perfect phase synchronisation

Since the phase of the system reference input is the integral of the frequency of the system reference input, an additional $1/s$ integrator term in the open-loop-system transfer function is required if zero steady-state error to a step change in phase, in the sense of Property (i) of Properties 2.1, is to be achieved. In turn, this additional $1/s$ integrator incurs a 90-degrees phase lag at all frequencies, including the open-loop system-gain crossover frequency. This destabilising phase lag needs to be compensated for by an additional phase lead term $(1+sT_1)$, such that the loop filter of Figure 2.3 becomes $(1+sT_1)/s(1+sT)$. In the region of the gain crossover frequency, the dominant closed-loop characteristic of the PLL will still, however, be that of a damped oscillator. The final PLL for perfect synchronisation (zero steady-state error) of the output signal phase is as depicted in Figure 2.4.

The phase-locked loop is a beautiful physically based example that should grace the pages of every control textbook and every history of control. The range of present-day applications of this ingenious control device in electronic systems, radio, television, computers, power electronics and communications is simply staggering. Most likely, the device is sitting next to you in your mobile telephone.

The physical phenomenon of weakly coupled oscillators (pendulum clocks, tuning forks, organ pipes and triode valves [tubes]), oscillating at nearly the same frequency, tending to synchronise so as to oscillate at the same frequency has been observed over the centuries. Appleton (1923) and Van der Pol (1926) each provide a thorough theoretical and experimental analysis of the phenomenon. There has also been considerable recent interest in the synchronisation phenomenon in various fields of science and engineering; see Francke, Pogromsky and Nijmeijer (2020). This phenomenon of natural oscillation synchronisation in physics can be likened to the inherent stability of dynamical systems through the dissipative effects embodied in Assumption 3.1 of Chapter 1. On the other hand, the PLL invented by Davidson two years after Appleton's (1923) paper, using a negative-feedback principle to meet a specified purpose, is an engineering triumph.

More generally, for larger interconnected systems such as power systems, the synchronisation phenomenon is one that arises from weakly coupled interconnected dynamical systems. There, the overall interconnected system dynamics evolve together over a slow time scale, while the local individual system dynamics have dissipated in a fast time scale, as surveyed by Saksena, O'Reilly and Kokotović (1984). Dynamically, the weak interconnections become strong in the slow time scale, and the system as a whole synchronises.

2.5 ANOTHER OUTLIER – THE POWER SYSTEM STABILISER (PSS)

This section examines another ingenious control device that has also been strangely neglected in the automatic control literature: the power system stabiliser (PSS). Again, as we shall see, the clue is in the name. It is recalled from Section 1.3 of Chapter 1 that the turbogenerator connected to an infinite bus is a two-input two-output electromechanical dynamical system where the two outputs are terminal voltage and shaft speed. The excitation control is a simple lag filter of the same form, as it happens, as the PLL loop filter in Figure 2.3. The customary controller $K(s)$ for this two-input two-output system is a diagonal controller that satisfies the fundamental conditions of Result 3.1 of Chapter 1, Result 2.1 and Assumption 3.1 of Chapter 1. For many years, this diagonal controller had worked fine in the industry for typical slow-acting excitation control.

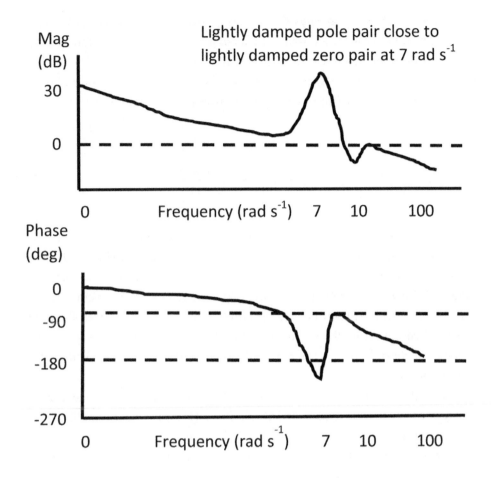

Figure 2.5 Awkward switch-back characteristic in turbogenerator excitation loop

Power system stabilisers were invented by US power utilities in the early 1960s to solve a pressing power system problem in more integrated systems, as described in Fadlalmoula et al. (1998). The increasing power system demands for faster-acting excitation systems in turbogenerators, so as to enable better post-fault system voltage recovery, entailed higher-excitation system gains and bandwidths than had been customary. This, in turn, caused an extremely awkward higher-frequency switch-back characteristic associated with lightly damped pole-zero pairs, first noted by Hamdan and Hughes (1977), to become excited, as the characteristic now lay close to the gain crossover frequency. How awkward this characteristic was is sketched in Figure 2.5. Moreover, the switch-back occurred at different frequencies, depending upon machine-loading conditions. What could be done to resolve this difficult issue and at what price?

In short, the PSS obviates this awkward switch-back characteristic, at a small price, to allow the desired benefits of a faster-acting excitation system. But how does the PSS device work? Well, the PSS consists of a cross-coupling loop transfer function $p(s)$ that feeds back the terminal voltage (output one) to the speed governor (input two). The transfer function $p(s)$ of the PSS takes the form

$$p(s) = k \frac{sT_w}{1+sT_w} \frac{1+sT_1}{1+sT_2} \qquad (2.3)$$

and consists of a gain, washout filter and phase-advance network, respectively. The PSS transfer function in (2.3) is tuned to provide a high-gain inverted notch characteristic that dominates the problematic switch-back characteristic in Figure 2.5 at 7 rads[-1]. In effect, at 7 rads[-1], but only at 7 rads[-1], the exciter largely ceases to control the terminal voltage output, but it is actually controlling the shaft speed output itself: a small price to pay. Otherwise, the turbogenerator excitation control proceeds as normal but over a newly extended bandwidth: a large benefit.

The PSS is installed worldwide in today's high-performance highly integrated electrical grids. Variants of PSS are also used to stabilise the latest power electronic devices as well. It deserves to be better known in the automatic control community. So as to reinforce the point, a device similar to the PSS could, in principle, be used to obviate other awkward high-frequency characteristics in control systems. Let us give it a name and call it a control system stabiliser. Another way of obviating awkward characteristics at a given frequency is through the judicious use of feedforward (Leithead and O'Reilly, 1993). Throughout, a more physical approach to control recommends itself.

2.6 COORDINATED DECENTRALISED CONTROL

Within the fundamental conditions of Result 3.1 of Chapter 1 that the plant outputs are linearly independent, and Result 2.1, in that a feedback controller exists, the two-input two-output turbogenerator connected to an infinite bus in Section 2.5 serves as an exemplar as to how one might proceed in the control of m-input m-output systems in general. In reality, that turbogenerator will be one of several generators in a much larger modern integrated power system with many other (closed) feedback loops out with the control of that power utility (Dudgeon et al., 2007; Dyśko, Leithead and O'Reilly, 2010; Xia, Dyśko and O'Reilly, 2015).

So, faced with the task of installing one such generator with associated controls, the engineer is likely to be guided by the experience and practices of their industry. In all likelihood, the control loops will be closed by the engineer, and the resulting performance assessed either first in simulation or more directly on the actual plant. Apart from further online tuning of the control gains from their manufacturer-recommended settings, the engineer may well be satisfied as to their closed-loop system performance. Indeed, Properties 2.1 of Section 2.2 confirm that all should be well at both low and high frequencies.

However, should the engineer have concern about the performance of the excitation loop, particularly in the mid-frequency range, they may seek to examine it by opening that loop, but keeping all other loops closed, as in the turbogenerator system of Section 2.5. In the reality of a multi-input multi-output interconnected system, which a modern power system is, the situation is exactly as depicted in Figure 2.6 (cf. Figure 1.2 of Chapter 1). In power systems, this integrated control framework is known as 'coordinated decentralised control', as in Dudgeon et al. (2007).

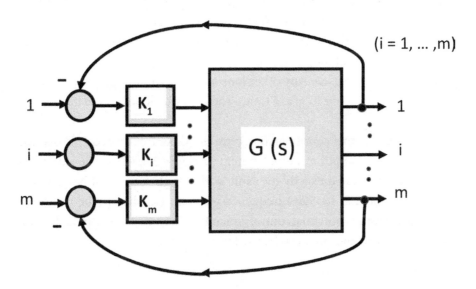

Figure 2.6 An m-*input* m-*output diagonal control system with all* m *feedback loops closed except for the* i-*th feedback loop, which is open*

Each (generator) controller is *coordinated* with all other (generator) controllers through the overall interconnected m-input m-output (power) system, as in Figure 2.6. Yet each (generator) controller can be treated as if it is a *decentralised* single-input single-output (SISO) controller connected only to its immediate (generator) subsystem, where classical SISO control pertains. No approximations are made, nor is any information from the overall m-input m-output interconnected (power) system lost.

The *coordinated decentralised control* approach described previously is physically based, completely general and applies to all other multi-input multi-output integrated control systems. An earlier version is to be found in O'Reilly and Leithead (1991, 1995) and Leithead and O'Reilly (1992b). Other areas where the approach has recently borne fruit[2] include power electronics (Guo et al., 2021; Pinares and Bongiorno, 2020; Wu et al., 2020; Ugalde-Loo, Acha and Licéaga-Castro, 2018), induction motors (Amézquita-Brooks et al., 2018), vehicle dynamics (Hermandez-Alcantara et al., 2018), gyroscopes (Hu and Gallacher, 2018) and wind generator systems (Wang et al., 2018). Earlier applications include helicopter flight control, as in Licéaga-Castro et al. (1995) and Dudgeon, Gribble and O'Reilly (1997).

Within this general framework, for the i-th open loop, one considers the i-th controller k_i as paired with the subsystem $g_{ii}(s)$, as if it were on its own. But that i-th open loop $k_i g_{ii}(s)$ is in fact interconnected with the whole of the remaining system and its myriad other closed-loop controls, as in Figure 2.6. This is exactly the situation for the turbogenerator of Section 2.5, where the excitation control loop is open, but the remaining one loop – the speed-governor loop – is closed. Any other loop available to the engineer, $i = 1, \dots ,$ m, may be opened – one loop at a time – with all other loops closed in the same fashion.

The beauty of this coordinated decentralised control approach is that the i-th loop of the m-input m-output system in Figure 2.6, when opened in this manner, can be treated exactly the same as the single-loop kg of Nyquist and Bode in Section 2.3. Its SISO frequency response can directly be measured and evaluated for any potential control difficulties. Should a suitable model $G(s)$ in Figure 2.6 be available, that can be used instead.

---

2   I am indebted to Professor J. Kocijan (2022), who has maintained for many years an online bibliography on the individual channel analysis and design framework.

Remark 6.1: Some readers may be unaccustomed to the control design of loops that depend upon other loop control designs, such as those described here in some detail for power systems. This, however, is the norm for many other industries, such as the automotive or aerospace sectors. There, controllers for this, that and the other – most proprietary to different companies – coexist and are embedded within the same overall m-input m-output interconnected system. Figure 2.6 and its related discussion on coordinated decentralised control nonetheless keeps the engineer from being overwhelmed by this level of modern industrial complexity. This issue is further explored in Section 2.7 on performance specifications and the process of interactive control design.

2.6.1 A DETAILED 'INSIDE LOOK' AT COORDINATED DECENTRALISED CONTROL

For a detailed inside look at coordinated decentralised control, let us return to where we started, with the fundamentals of Definition 3.1 of Chapter 1 and Result 2.1 of Section 2.2. By Mirsky (1955), the determinant $Det\ G(s)$ of the m x m plant matrix $G(s)$ of Figure 2.6 may be expanded in terms of the elements $g_{ij}(s)$ of the i-th row according to

$$Det\ G(s) = \sum_{j=1}^{m} g_{ij}G_{ij} \quad (i=1,\ldots,m) \tag{2.4}$$

where $G_{ij}(s)$ is the cofactor of the transfer-function element $g_{ij}(s)$. Separating out the 'diagonal' term of (2.4), one has that

$$Det\ G(s) = g_{ii}G_{ii} + \sum_{j=1}^{m} g_{ij}G_{ij} \quad (i=1,\ldots,m; i \neq j) \tag{2.5}$$

Then, equation (2.5) may be rearranged to obtain the following expression:

$$g_{ii} + \sum_{j=1}^{m} g_{ij}G_{ij}\ /\ G_{ii} \quad (i=1,\ldots,m; i \neq j) \tag{2.6}$$

Recall that Condition (i) of Result 3.1 of Chapter 1 requires that the m x m transfer-function matrix $G(s)$ in Figure 2.6 be non-singular. This is equivalent to requiring that each of the scalar expressions in (2.4) to (2.6) be non-zero. Indeed, this non-zero condition on the scalar transmittance of (2.6) is similar to that on the scalar transfer function $g(s)$ in the single-loop case of Section 2.3.

On recalling that the i-th controller k_i is paired with the diagonal transfer-function element $g_{ii}(s)$ in Figure 2.6, the second term of the scalar transmittance in (2.6) appears to take some additional account of the interactions from the closed loops of Figure 2.6. Property (i) of Properties 2.1 in Section 2.2 lends support to this notion in that, for high gain at low frequencies, the closed-loop-system transfer-function matrix in Figure 2.6 will be an identity matrix except for the i-th loop, which is open.

Within the fundamental conditions of Result 3.1 of Chapter 1 and Result 2.1 of Section 2.2, the *actual* open-loop transmittance for the i-th loop in Figure 2.6, with controller gains not necessarily high, is obtained as follows. Just as we had previously proceeded with $G(s)$ in Condition (i) of Result 3.1 of Chapter 1 to obtain the scalar transmittance in (2.6), let us now work with the return-difference matrix of Result 2.1 and Figure 2.1, by way of the related $m \times m$ matrix $\bar{G}(s)$, defined as

$$\bar{G}(s) = K(s)^{-1} + G(s) \qquad (2.7)$$

The matrix $K(s)^{-1} + G(s)$ in (2.7), representing the return-difference matrix of Result 2.1, can be thought of as a translation of the plant matrix $G(s)$. In the scalar case, this amounts to a shift of origin to $(-1/k, 0)$, as observed in the polar plot of $g(s)$, such as that presented in Power and Simpson (1978).

Remark 6.2: As in Remark 2.2, the $m \times m$ matrix $\bar{G}(s)$ in (2.7) preserves the independence of the m inputs and the independence of the m outputs, as they originally arose with the $m \times m$ plant matrix $G(s)$ in the feedback arrangement.

Alternatively, the $m \times m$ matrix $\bar{G}(s)$ of (2.7) is defined as

$$\bar{G}(s) = H(s)^{-1} G(s) \qquad (2.8)$$

where the $m \times m$ closed-loop system matrix $H(s)$ is defined as

$$H(s) \triangleq GK[I + GK]^{-1} \qquad (2.9)$$

By Result 2.1, it is required that the $m \times m$ matrix $\bar{G}(s)$ in (2.7) or (2.8) be non-singular, or that $Det\ \bar{G}(s)$ be non-zero. Therefore, we may proceed directly in analogy with equations (2.4) to (2.6) to obtain the *actual* i-th open-loop scalar transmittance $\hat{g}_i(s)$ of Figure 2.6 (Leithead and O'Reilly, 1992b), described in our notation by

$$\hat{g}_i(s) = g_{ii} + \sum_{j=1}^{m} \overline{g}_{ij} \overline{G}_{ij} / \overline{G}_{ii} \qquad (i = 1,\ldots,m; i \neq j) \tag{2.10}$$

where $\overline{G}_{ij}$ is the cofactor of the transfer-function element $\overline{g}_{ij}(s)$ of the $m \times m$ matrix $\overline{G}(s)$ of (2.7) or (2.8). The matrix $\overline{G}(s)$ in (2.7) can be written out in full by noting that the diagonal elements of $\overline{G}(s)$, for diagonal controller $K(s)$, are of the form

$$k_i^{-1} + g_{ii} = g_{ii}(1 + k_i g_{ii}) / (k_i g_{ii}) = g_{ii} / h_i \qquad (i = 1,\ldots,m) \tag{2.11}$$

where the i-th closed-loop subsystem transfer function, $h_i(s)$, is

$$h_i(s) \triangleq k_i g_{ii} / (1 + k_i g_{ii}) \tag{2.12}$$

Hence, we have $\overline{G}(s)$ given by

$$\overline{G}(s) = \begin{bmatrix} g_{11} / h_1 & g_{12} & \cdots & g_{1m} \\ g_{21} & g_{22} / h_2 & \cdots & g_{2m} \\ \vdots & & \ddots & \vdots \\ g_{m1} & g_{m2} & \cdots & g_{mm} / h_m \end{bmatrix} \tag{2.13}$$

Remark 6.3: In accordance with Remark 3.3 of Chapter 1 and Remark 2.3, so as to avoid an ill-defined feedback system (and ill-defined feedback properties), it is also necessary that both the i-th scalar transmittance (2.6) and the i-th open-loop scalar transmittance $\hat{g}_i(s)$ in (2.10) be not near zero either. Put another way, the polar plots of (2.6) and $\hat{g}_i(s)$ in (2.10) should not be close to the origin of the complex plane.

The open-loop transmittance $\hat{g}_i(s)$ in (2.10) is the *actual* open-loop transmittance for the i-th loop that is paired with the i-th controller $k_i(s)$ in Figure 2.6. It takes complete account of the rest of the plant $G(s)$ in Figure 2.6 with the remaining $m - 1$ feedback loops closed; no interaction information is lost. There will, however, be an interaction disturbance affecting the i-th output y_i, via signal transmission in Figure 2.6 from the remaining $m - 1$ reference signals r_j (j ≠ i); see Leithead and O'Reilly (1992b). This disturbance is rejected in the usual fashion by the controller gain $k_i(s)$ in the sense of Property (i) of Properties 2.1 in Section 2.2. When the remaining $m - 1$ controller gains are high ($h_j = 1$; j ≠ i), the scalar transmittance (2.10) reduces to that of (2.6), as previously intimated.

For the first open-loop transmittance $\hat{g}_1(s)$ to which the controller $k_1(s)$ in Figure 2.6 is paired, we have from (2.10)

$$\hat{g}_1(s) = g_{11} + \sum_{j=1}^{m} \bar{g}_{1j} \bar{G}_{1j} / \bar{G}_{11} \qquad (j \neq 1)$$

$$= g_{11} + \frac{\begin{vmatrix} 0 & g_{12} & \cdots & g_{1m} \\ g_{21} & g_{22}/h_2 & \cdots & g_{2m} \\ \vdots & & \ddots & \vdots \\ g_{m1} & g_{m2} & \cdots & g_{mm}/h_m \end{vmatrix}}{\begin{vmatrix} g_{22}/h_2 & \cdots & g_{2m} \\ \vdots & \ddots & \vdots \\ g_{m2} & \cdots & g_{mm}/h_m \end{vmatrix}} \qquad (2.14)$$

where $|X|$ denotes the determinant of a matrix X.

Recall that Result 2.1 of Section 2.2 requires that the system return-difference matrix be non-singular. This is equivalent to requiring that the i-th scalar transmittance $\hat{g}_i(s)$ of (2.10) be non-zero. Were Result 2.1 to be violated with the return-difference matrix singular, the scalar transmittance $\hat{g}_i(s)$ of (2.10) would be zero. In that singular case, the local subsystem $g_{ii}(s)$ in (2.10) is *completely negated* by the second system-wide interaction term $\sum_{j=1}^{m} \bar{g}_{ij} \bar{G}_{ij} / \bar{G}_{ii}$, $(j \neq i)$.

We may now summarise this in the following remark.

Remark 6.4: The decomposition of the m-input m-output system, as represented by the return-difference matrix of Result 2.1, into m SISO transmittances (2.10) is exact. These m SISO transmittances $\hat{g}_i(s)$ of (2.10) capture all the structure of the original m-input m-output system. Consequently, the ability or otherwise to meet performance objectives and specifications on each of the m outputs rests solely with each of the m SISO transmittances paired with its respective scalar controller. This is little different to SISO classical control, as shall be explored further in Section 2.7.

Remark 6.5: Historically, different authors have imposed various restrictions on the i-th scalar transmittance $\hat{g}_i(s)$ of (2.10) so as to reduce dependence on the other controllers $k_j(s)$ in the second system-wide interaction term $\sum_{j=1}^{m} \bar{g}_{ij} \bar{G}_{ij} / \bar{G}_{ii}, (j \neq i)$. Rosenbrock (1969, 1974) requires that the first local diagonal term $g_{ii}(s)$ be dominant relative to the second system-wide interaction term in (2.10). (The non-singularity conditions of Result 3.1 of Chapter 1 and Result 2.1 are replaced by the much stronger conditions of diagonal dominance.) Horowitz (1979) goes even further in assuming infinitely high controller gains, thereby entirely eliminating the dependence on the other controllers $k_j(s)$ in the second interaction term of (2.10) in the sense of Properties 2.1. This assumption is relaxed to a wide bandwidth separation requirement between loops in Horowitz (1982). O'Reilly and Leithead (1991) impose a less conservative loop bandwidth separation requirement. The coordinated decentralised control described here imposes no restrictions; the i-th scalar transmittance $\hat{g}_i(s)$ of (2.10) is what it is, controllers and all.

2.6.2 COORDINATED DECENTRALISED CONTROL SUMMARY

Well, we have had a good look at the 'engine' of coordinated decentralised control in Section 2.6.1. This concluding section summarises all that the driver needs to know in order to drive the vehicle away (under the fundamental conditions of Result 3.1 of Chapter 1 that the plant outputs are linearly independent, and Result 2.1 of Section 2.2 that a feedback controller exists, of course).

Consider the m-input m-output system represented by Figure 2.6 where the i-th feedback loop is open, but all other feedback loops are closed. That i-th open-loop single-input single-output (SISO) frequency response can directly be measured and evaluated using classical SISO control for any potential control difficulties. Any other loop available to the engineer, $i = 1, \ldots, m$, may be opened and evaluated – one loop at a time – with all other loops closed in the same way. Should a suitable model $G(s)$ in Figure 2.6 be available, the SISO transmittance $\hat{g}_i(s)$ of (2.10) can be used instead. It is that simple. And all from earlier physical arguments.

We are now in a position to summarise these results by way of the following result.

Result 6.1: Under the conditions of Result 3.1 of Chapter 1, namely the non-singularity of the square $m \times m$ plant matrix $G(s)$ and system causality, unity negative-feedback scalar controllers k_i, $i = 1, \ldots, m$, as depicted in Figure 2.6, with scalar loop-gains $k_i(s)\hat{g}_i(s)$ exist if and only if the open-loop scalar transmittances given by

$$\hat{g}_i(s) = g_{ii} + \sum_{j=1}^{m} \overline{g}_{ij} \overline{G}_{ij} / \overline{G}_{ii} \qquad \left(i = 1, \ldots, m; i \neq j\right) \tag{2.15}$$

are each non-zero, where $\overline{G}_{ij}$ is the cofactor of the transfer-function element $\overline{g}_{ij}(s)$ of the $m \times m$ matrix $\overline{G}(s)$ defined as

$$\overline{G}(s) = K(s)^{-1} + G(s) \tag{2.16}$$

Remark 6.6: Result 2.1 of Section 2.2 and Result 6.1 are entirely equivalent. The former Result 2.1 is more useful as a compact statement of the existence of a negative-feedback controller; the latter, Result 6.1, is more useful for direct industrial application. The facts of a possible non-diagonal $m \times m$ controller matrix $K(s)$ in Result 2.1 and a diagonal $m \times m$ controller matrix $K(s)$ in Result 6.1 are reconciled by noting that any non-diagonal controller matrix $K(s)$ can be rewritten as $K_{comp}(s)K_{diag}(s)$, where $K_{comp}(s)$ is a non-diagonal plant pre-compensator matrix and $K_{diag}(s)$ is a diagonal controller matrix.

Result 6.1 can be thought of as a highly useful rearrangement of the fundamental Result 2.1 to suit the purposes of coordinated decentralised control.

Remark 6.7: A rereading of relations (2.7) to (2.10) reveals that the open-loop transmittance (2.10) in Section 2.6.1 does not require the square controller matrix $K(s)$ to be diagonal; it may be non-diagonal. This means that the plant diagonal elements $g_{ii}(s)$ can be separated out as before, but paired with the controller elements $k_{ii}(s)$. It thereby provides a direct assessment of non-diagonal controllers from other designs; for instance, eigenstructure assignment (Fahmy and O'Reilly, 1988). Simply plug the non-diagonal controller in and open any feedback loop of interest as in Figure 2.6.

2.7 PERFORMANCE SPECIFICATIONS AND THE PROCESS OF INTERACTIVE CONTROL DESIGN

Our attention now turns full circle to control objectives and the purpose of automatic control. Those control objectives, if they are to be achieved, will need to be fleshed out in the form of suitable performance specifications. The control objectives are then only achieved when the performance specifications are satisfactorily met by control systems design. That control systems design is a process during which performance specifications are to be increasingly met despite dynamical constraints to the contrary in the uncontrolled system.

Notwithstanding Assumption 3.1 of Chapter 1, one such dynamical constraint might be that the system is unstable; worse still, the system might be non-minimum phase. In these infrequent cases of unstable or non-minimum-phase open-loop systems, a progressive account of what can or cannot be achieved is presented in Leithead and O'Reilly (1991), using Nyquist encirclement arguments. There may be other physical constraints particular to the uncontrolled system.

Taking the coordinated decentralised control framework of Section 2.6 as the most general case, it is therefore helpful to encapsulate our notion of performance specifications, as they might be realised, in the following definition.

Definition 7.1: Under the fundamental conditions of Result 3.1 in Chapter 1 that the m plant outputs are linearly independent, and Result 2.1 in Section 2.2 that a feedback controller exists, the performance specifications on each of the m individual outputs $y_i(s)$, $i = 1,\ldots,m$, refer to performance specifications that are attainable in the face of the inevitable dynamical constraints on the SISO loop transmittances $\hat{g}_i(s)$ in (2.15) of Result 6.1. As with the classical single-loop design in Section 2.3, such specifications are either known a priori or they may need modification as the design process evolves.

It is the case, within the coordinated decentralised control framework of Section 2.6, that the whole burden of m-input m-output control falls fairly and squarely on the m SISO transmittances $\hat{g}_i(s)$ of (2.15) of Result 6.1, each with their associated scalar controller $k_i(s)$. Individually and collectively, all the m SISO controller designs will proceed in an interactive fashion, trading performance specifications with transmittance constraints, consistent with Definition 7.1. This bears all the hallmarks of SISO classical control design.

Within the experience of the engineer, initial approximations for the other controllers $k_j(s)(j \neq i)$ embedded within the second interaction term of the transmittances $\hat{g}_i(s)$ of (2.15) in Result 6.1 are themselves discarded quickly in favour of the latest design values as the interactive controller design process unfolds. The engineer, of course, may not enjoy these extra degrees of design freedom, since those other controller designs embedded within the second interaction term of the transmittances $\hat{g}_i(s)$ of (2.15) in Result 6.1 may have been preset by other control vendors.

In any case, the frequency response of any critical open loop, with all other loops closed, as shown in Figure 2.6 can be investigated experimentally. Like the classical single-loop design of Section 2.3, the process of interactive design described here is a profoundly human activity, and engineers are extremely good at it.

As with engineering design in general, the specifications may be too demanding, the constraints too severe, and the engineer too inexperienced. Then, if the overall control objectives are to be achieved, there is nothing for it but to start again with more relaxed specifications and a more experienced engineer, since the design constraints are unlikely to be weakened. Inevitably, system constraints translate into lower closed-loop system bandwidth, and hence lower closed-loop-system performance, than one would have liked.

This process of interactive control system design is summarised with reference to Result 6.1 in the following step-by-step design procedure.

Design Procedure 7.1: With an eye to the performance specifications of Definition 7.1 and a priori knowledge of controllers and system:

(0) Make a first guess for controllers $k_1,\ldots,k_m$.
(1) For loop 1, redesign controller k_1 based on $\hat{g}_1(k_2,\ldots,k_m)$ of (2.15). Then update k_1 in the pool of controllers $k_1,\ldots,k_m$ of Step (0).
(2) For loop 2, redesign controller k_2 based on $\hat{g}_2(k_1,k_3,\ldots,k_m)$ of (2.15). Then update k_2 in the pool of controllers $k_1,\ldots,k_m$ of Step (0).

...

(m) For loop m, redesign controller k_m based on $\hat{g}_m(k_1,\ldots,k_{m-1})$ of (2.15). Then update k_m in the pool of controllers $k_1,\ldots,k_m$ of Step (0).
Repeat Steps (1) to (m) as necessary.

At each successive step of Design Procedure 7.1, the performance specifications of Definition 7.1 are increasingly pulling the SISO controllers $k_1, \ldots, k_m$ towards what they should be in order to meet said specifications. So, as the step-by-step design procedure unfolds, the initial choices for the SISO controllers $k_1, \ldots, k_m$ in Step (0) become less important. This is as it should be. It accords with industrial practice. The next example illustrates these points.

Example 7.1: Consider the five-machine-equivalent Brazilian network of Dyśko, Leithead and O'Reilly (2010). Two power system stabilisers – PSS_1 and PSS_2, in the form of equation (2.3) in Section 2.5 – are to be designed. For the resulting strongly coupled two-input two-output PSS system, PSS_1 and PSS_2 are designed according to Design Procedure 7.1.

(0) With an eye to the specifications and the structure of the two PSSs as in (2.3), make a first guess for controllers PSS_1 and PSS_2.

(1) For PSS loop 1, redesign controller PSS_1 based on $\hat{g}_1\left(PSS_2\right)$ of (2.15). Then update PSS_1 in the pool of controllers PSS_1 and PSS_2 of Step (0).

(2) For PSS loop 2, redesign controller PSS_2 based on $\hat{g}_2\left(PSS_1\right)$ of (2.15). Then update PSS_2 in the pool of controllers PSS_1 and PSS_2 of Step (0).

Repeat Steps (1) and (2) as necessary.

Remark 7.1: The actual design procedure of controllers PSS_1 and PSS_2 executed in Dyśko, Leithead and O'Reilly (2010) for Example 7.1 consists of just three steps – (0), (1) and (2) – without repetition. In Step (0), the PSS_2 gain was chosen to be infinite over all frequencies; this was a convenient initial choice, but one extremely poor at capturing the characteristics of the PSS in (2.3) of Section 2.5, which is only high gain at a specific frequency and is otherwise rolled off at all other frequencies. Nonetheless, the two subsequent steps, (1) and (2), resulted in designs for PSS_1 and PSS_2 that discounted this initial choice of PSS_2 in Step (0) to meet the required design specifications comfortably.

Remark 7.2: A loop bandwidth separation restriction is invoked in O'Reilly and Leithead (1991). Close examination reveals, however, that this initial restriction is quietly discarded in subsequent design steps, along the lines of Design Procedure 7.1.

And finally, just as we previously turned full circle to control objectives at the start of this section, let us now return to the *m*-input *m*-output design process itself, as found at the start of Section 2.6. There, it was described how an industrial engineer might execute the control design directly with respect to Figure 2.6. That process is summarised in the next step-by-step design procedure.

Design Procedure 7.2: With an eye to the performance specifications of Definition 7.1, Figure 2.6, and a priori knowledge of controllers and system:

(1) Open loop 1, design controller k_1. Close loop 1.
(2) Open loop 2, design controller k_2. Close loop 2.

  ...

(*m*) Open loop *m*, design controller k_m. Close loop *m*.
Repeat Steps (1) to (*m*) as necessary.

In Design Procedure 7.2, the mantra with respect to Figure 2.6 is open, design, close. It is consistent with industrial practice. As in Remark 6.1, the engineer faced with troublesome control loops in a complex interconnected system may scarcely be aware of the functions, let alone the details of other control loops designed by other companies. Nonetheless, for those particular control loops of interest, Design Procedure 7.2's mantra of 'open, design, close' should prove to be invaluable. Should that engineer chance upon this chapter, they may well be seen to smile and murmur, "Well, what do you know?"

2.8   CONCLUSIONS

This chapter is an affirmation of the greatness of automatic control and the need for a more physical approach towards it. The origins of the chapter go back a long way; they have been fed from many years of research in control engineering, power systems and power electronics. The author has also been most fortunate in being privy to the thinking of some of the greatest exponents in these fields.

Aside from physical realisability, the fundamental underlying issue of automatic control is *singularity* – and how to avoid it. This is as true for control systems with one input and one output as it is for systems with several inputs and outputs. The singularity issue starts with the uncontrolled linear system or plant description, as in Chapter 1. It then extends to both the open-loop and closed-loop control system by way of the return-difference matrix – it also must not be singular. All of this is implicit in the development of frequency response methods of automatic control (MacFarlane, 1979). In brief, what is shown is that automatic control is an adjunct of physical dynamical systems, as defined in Chapter 1.

Recognition of this fundamental singularity issue, recast more positively as an essential non-singularity issue, allows for the interpretation and derivation of automatic control in a much more direct and physically appealing manner. It also gives rise to a new framework for the coordinated decentralised control of multi-input multi-output systems, not so very different from classical control, as discussed in Section 2.6. This new coordinated decentralised control framework effectively cuts through the complexities of modern industrial control practice. The original idea behind Section 2.7, on performance specifications and the process of control design, arose many years ago from the work of Polanyi (1973) on engineering purpose and principles of operation. The current development takes its cue from O'Reilly and Leithead (1991).

The idea of Condition (i) of Definition 3.1 of Chapter 1 is enshrined in the opening control requirements of O'Reilly and Leithead (1991): to each plant reference input there should be one, and only one, corresponding plant output. Condition (ii) of Definition 3.1 of Chapter 1 on physical realisability is, of course, fundamental to dynamical systems in general. The original papers of Nyquist (1932) and Bode (1945) on the stability and performance of feedback amplifiers, pioneering though they were, pay scant attention to dynamical systems, save the assumption of causality in Nyquist (1932). The connection to earlier work on dynamical systems is, however, obvious enough (MacFarlane, 1979; Power and Simpson, 1978). What Chapter 1 and Chapter 2 establish is how fundamental the dynamical system concepts of both causality and independence of system outputs (non-singularity) are to automatic control, regardless of the numbers of system inputs and system outputs under consideration.

In a sense, by adopting a more physical approach to automatic control, the chapter starts and finishes with Definition 3.1 of Chapter 1. This definition allows us to place automatic control firmly within the tradition of physical dynamical systems. All other results in Section 2.2 follow immediately from it. Indeed, all the remaining sections are a reinterpretation of existing historical results in the context of the more physical approach represented by Definition

3.1 of Chapter 1. The shift in thinking may appear subtle, but it amounts to nothing less than a total repositioning of automatic control to be all of a piece. This has profound benefits. One immediate benefit is that the basic relations of automatic control achieve their full significance when viewed in this more physical way. Another benefit is to place system stability more within overall control objectives. After all, Assumption 3.1 of Chapter 1 reminds us that, for physical reasons, most systems are stable anyway, without control. It is just that, in pursuing control performance, one needs to be aware that increased performance comes at the expense of system stability.

Yet another significant benefit of adopting a more physical approach to automatic control is that the technical challenges represented by a particular control problem are of immediate interest to *both* the control theorist and the practising engineer. The engineer with a technical challenge will have need of the specialist. The specialist, their interest aroused by the technical possibilities of the immediate problem, will have their awareness of its wider engineering significance enhanced. Both are engaged in a mutually rewarding activity. The old theory-applications dichotomy simply disappears. Section 2.3 on the historical negative-feedback amplifier, in particular, bears this out.

As a further example, the aerospace industry, like any other industry, has its own particular requirements. Among them, flight control places great importance on transparent and tested methods of design. It is after all a safety-critical environment. In addition to classical control methods of the type described in this chapter, modal methods of control are extensively used. Again, a physical approach recommends itself. One such physically based modal control approach is *parametric eigenstructure assignment* (Fahmy and O'Reilly, 1988). This parametric approach differs radically from the classical eigenvalue-eigenvector approach of numerical linear algebra in that it directly exploits the well-known (A, B, C) structure of state-space models to parameterise the eigenvectors. Together with the eigenvalues, these additional physical tuning parameters are at the direct disposal of the design engineer to tune the system response as they see fit.

But such a physical approach also holds great promise in providing the needed context and motivation for future discoveries and inventions. For example, deep physical knowledge of the dynamics of the system to be controlled can be used to advantage in devising novel high-performance control systems. Such is the case with the car emergency collision avoidance manoeuvre of Bevan, Gollee and O'Reilly (2007), where a physically based composite controller design accomplished astonishing performance over a wide envelope of car speeds.

Automatic control is an engineering discipline. By engineering, we have in mind the engineer as *ingénieur* or inventor, in the European sense of the word. As such, automatic control has its own *modus operandi*. That mode of operation is to satisfy overall control objectives – the overarching purpose of the discipline. This is abundantly clear with the negative-feedback amplifier, phase-locked loop and power system stabiliser of Sections 2.3 to 2.5. Each harnesses physical laws within a feedback mode of operation to meet a clearly defined engineering need or purpose. New technical developments thereby arise.

Moreover, this physical engineering spirit is alive and well today, if one looks for it. The present chapter, with its shameless self-referencing, is based upon the author's own experience. This very same spirit is to be found in other areas of engineering, such as biomedical, environmental, renewable energy and rehabilitation engineering, where automatic control features more often that one might suppose. There are new interesting problems out there. New technical developments in automatic control will ensue. Surely, they must benefit from the adoption of a more physical approach, such as that described in this chapter.

EPILOGUE TO PART I

In the preceding discussion, an emerging philosophy of engineering – focused, as it were, on dynamical systems and automatic control – has been sketched. Unlike its scientific counterpart (Polanyi, 1973), it is in embryonic form. It could be developed. There is certainly interest (Dias, 2019; Mitcham, 2019). At the heart of this engineering philosophy – what is engineering – is likely to be a physical narrative as to purpose and principles of operation. That physical narrative, as well as providing a necessary philosophy for engineering, may in turn yield important new results and new research directions, as it did so generously here in the case of dynamical systems and automatic control, where its need was so acute.

REFERENCES FOR PART I

Amézquita-Brooks, L. A., Ugalde-Loo, C. E., Licéaga-Castro, E. and Licéaga-Castro, J. (2018). In-depth cross-coupling analysis in high-performance induction motor control. *Journal of the Franklin Institute, 355,* 2142–2178.

Andronov, A. A., Vitt, A. A., and Khaikin, S. E. (2011). *Theory of Oscillators.* Dover, New York.

Appleton, E. V. (1923). The automatic synchronization of triode oscillators. *Proceedings of the Cambridge Philosophical Society, 21,* 231–248.

Birkhoff, G. D. (1927). *Dynamical Systems.* American Mathematical Society, Providence RI.

Black, H. S. (1934). Stabilized feedback amplifiers. *Bell System Technical Journal, 13,* 1–18.

Black, H. S. (1977). Inventing the negative-feedback amplifier. *IEEE Spectrum, 14,* 54–60.

Bevan, G., Gollee, H. and O'Reilly, J. (2007). Automatic collision avoidance manoeuvre for a passenger car. *International Journal of Control, 80,* 1751–1762.

Bode, H. W. (1940). Relations between attenuation and phase in feedback amplifier design. *Bell System Technical Journal, 19,* 421–454.

Chua, L. O. (1969). *Nonlinear Network Theory.* McGraw-Hill, New York.

Clarke, D. W. and Park, J. W. (2003). Phase-locked loops for plant tuning and monitoring. *IEE Proceedings Control Theory and Applications, 150,* 155–169.

Dias, P. (2019). *Philosophy of Engineering: Practice context, ethics, models, failure.* Springer-Verlag, Singapore.

Dudgeon, G. J. W., Gribble, J. J. and O'Reilly, J. (1997). Individual channel analysis and helicopter flight control in moderate and large amplitude manoeuvres. *Control Engineering Practice, 5,* 33–38.

Dudgeon, G. J. W., Leithead, W. E., Dyśko, A., O'Reilly, J. and McDonald, J. R. (2007). The effective role of AVR and PSS in power systems: Frequency response analysis. *IEEE Transactions on Power Systems, 22,* 1986–1994.

Dyśko, A., Leithead, W. E. and O'Reilly, J. (2010). Enhanced power system stability by coordinated PSS design. *IEEE Transactions on Power Systems*, *25*, 413–422.

Fahmy, M. M. and O'Reilly, J. (1988). Parametric eigenstructure assignment by output-feedback control: The case of multiple eigenvalues. *International Journal of Control*, *48*, 1519–1535.

Fadlalmoula, Z., Robertson, S. S., O'Reilly, J. and Leithead, W. E. (1998). Individual channel analysis of the turbogenerator with a power system stabiliser. *International Journal of Control*, *69*, 175–202.

Francke, M., Pogromsky, A. and Nijmeijer, H. (2020). Huygen's clocks: 'sympathy' and resonance. *International Journal of Control*, *93*, 274–281.

Guo, C., Yang, S., Liu, W., Zhao, C. and Hu, J. (2021). Small-signal stability enhancement approach for VSC-HVDC system under weak AC grid conditions based on single-input single-output transfer function model. *IEEE Transactions on Power Delivery*, *36*, 1313–1323.

Hamdan, A. M. and Hughes, F. M. (1977). Analysis and design of power system stabiliser. *International Journal of Control*, *26*, 769–782.

Hermandez-Alcantara, D., Amézquita-Brooks, L., Morales-Menendez, R., Sename, O. and Dugard, L. (2018). The cross-coupling of lateral-longitudinal vehicle dynamics: Towards decentralized fault-tolerant control schemes. *Mechatronics*, *50*, 377–393.

Horowitz, I. (1979). Quantitative synthesis of uncertain multiple input-output feedback systems. *International Journal of Control*, *30*, 81–106.

Horowitz, I. (1982). Improved design technique for uncertain multiple input-output feedback systems. *International Journal of Control*, *36*, 977–988.

Hu, Z. and Gallacher, B. (2018). A mode-matched force-rebalance control for a MEMS vibratory gyroscope. *Sensors and Actuators A: Physical*, *273*, 1–11.

Kalman, R. E. (1963). Mathematical description of linear dynamical systems. *SIAM Journal of Control*, *1*, 152–192.

Kocijan, J. (2022). *Individual Channel Analysis and Design*. Available at: http://dsc.ijs.si/jus.kocijan/ICAD/

Lanczos, C. (1996). *Linear Differential Operators*. SIAM Classics in Applied Mathematics, Philadelphia.

Leithead, W. E. and O'Reilly, J. (1991). Uncertain SISO systems with fixed stable minimum-phase controllers: Relationship of closed-loop systems to plant RHP poles and zeros. *International Journal of Control*, *53*, 771–798.

Leithead, W. E. and O'Reilly, J. (1992a). Performance issues in the individual channel design of 2-input 2-output systems, Part 3: Non-diagonal control and related issues. *International Journal of Control, 55*, 265–312.

Leithead, W. E. and O'Reilly, J. (1992b). *m*-Input *m*-output feedback control by individual channel design, Part 1: Structural issues. *International Journal of Control, 56*, 1347–1397.

Leithead, W. E. and O'Reilly, J. (1993). New roles for feedforward in multivariable control by individual channel design. *International Journal of Control, 57*, 1357–1386.

Licéaga-Castro, J., Verde, C., O'Reilly, J. and Leithead, W. E. (1995). Helicopter flight control by individual channel design. *Proceedings of the Institution of Electrical Engineers, 142*, 58–72.

Lyapunov, A. M. (1992). *The General Problem of the Stability of Motion*, translated by A. T. Fuller. Taylor & Francis, London.

MacFarlane, A. G. J. (1979). The development of frequency response methods in automatic control. *IEEE Transactions on Automatic Control, 24*, 250–265.

Maxwell, J. C. (1868). On governors. *Proceedings of the Royal Society of London, 16*, 270–283.

Mayr, O. (1971a). Victorian physicists and speed regulation: An encounter between science and technology. *Notes and Records of the Royal Society of London, 26*, 205–226.

Mayr, O. (1971b). Maxwell and the origins of cybernetics. *Isis, 62*, 425–444.

Mindell, D. A. (2000). Opening Black's box: Rethinking feedback's myth of origin. *Technology and Culture, 41*, 405–434.

Mirsky, L. (1955). *An Introduction to Linear Algebra.* Oxford.

Mitcham, C. (2019). *Steps Towards a Philosophy of Engineering: Historico-Philosophical and Critical Essays.* Rowman & Littlefield International, Lanham Maryland.

Muthuswamy, B. and Banerjee, S. (2019). *Introduction to Nonlinear Circuits and Networks.* Springer, Berlin.

Nayfeh, A. H. and Mook, D. T. (1979). *Nonlinear Oscillations.* John Wiley & Sons, New Jersey.

Nyquist, H. (1932). Regeneration theory. *Bell System Technical Journal, 11*, 126–147.

O'Reilly, J. (1983). *Observers for Linear Systems.* Academic Press, London.

O'Reilly, J. and Leithead, W. E. (1991). Multivariable feedback control by individual channel design. *International Journal of Control, 54*, 1–46.

O'Reilly, J. and Leithead, W. E. (1995). Frequency-domain approaches to multivariable feedback control systems design: an assessment by individual channel design for 2-input 2-output systems. *Control Theory and Advanced Technology, 10,* 1913–1940.

Pinares, G. and Bongiorno, M. (2020). Independent channel approach for stability analysis of grid-connected converters. *Electric Power System Research, 189, 106774.*

Poincaré, H. J. (1892-1899). *Les méthodes nouvelles de la mécanique céleste,* Volumes 1-3. Gauthiers-Villars, Paris.

Polanyi, M. (1973). *Personal Knowledge.* Routledge & Kegan Paul, London.

Power, H. M. and Simpson, R. J. (1978). *Introduction to Dynamics and Control.* McGraw-Hill, Maidenhead England.

Rosenbrock, H. H. (1969). Design of multivariable control systems using the inverse Nyquist array. *IEE Proceedings, 116,* 1929–1936.

Rosenbrock, H. H. (1974). *Computer-Aided Control Systems Design.* Academic Press, London.

Saksena, V.R., O'Reilly, J. and Kokotović, P.V. (1984). Singular perturbation and time-scale methods in control theory: Survey 1976–1983. *Automatica, 20,* 273–293.

Smale, S. (1967). Differentiable dynamical systems. *Bulletin of the American Mathematical Society, 73,* 747–817.

Ugalde-Loo, C. E., Acha, E. and Licéaga-Castro, E. (2018). Analysis of the damping characteristics of two power electronics-based devices using 'individual channel analysis and design'. *Applied Mathematical Modelling, 59,* 527–545.

Van der Pol, B. (1926). On "relaxation-oscillations". *Philosophical Magazine, 2,* 978–992.

Wang, D., Liang, L., Hu, J., Chang, N., and Hou, Y. (2018). Analysis of low-frequency stability in grid-tied DFIGs by nonminimum phase zero identification. *IEEE Transactions on Energy Conversion, 33,* 716–729.

Wu, W., Liu, J., Li, Y. and Blaabjerg, F. (2020). Individual channel design design-based precise analysis and design for three-phase grid grid-tied inverter with LCLLCL-filter under unbalanced grid impedance. *IEEE Transactions on Power Electronics, 35,* 5381–5396.

Xia, J., Dyśko, A. and O'Reilly, J. (2015). Future stability challenges for the UK network with high wind penetration levels. *IET Proceedings on Generation, Transmission & Distribution, 9,* 1160–1167.

Zivanovic, M. D. and Vukobratovic, M. (2006). *Multi-arm Cooperating Robots.* Springer, Berlin.

PART 2

SINGULAR PERTURBATIONS AND TWO-FREQUENCY-SCALE SYSTEMS

PREFACE TO PART 2

Part 2 of this book is the unlikely outcome of a recent return to the world of singular perturbations, initial value problems and approximate models for dynamical systems. And that after an absence of some thirty years spent mainly in power systems, power electronics and wind energy systems research; all dynamical systems, of course.

Regarding singular perturbations, the key link to the past is the book *Singular Perturbation Methods in Control*, originally published in 1986, and republished as Kokotović, Khalil and O'Reilly (1999). But this is no nostalgic revisit – time becomes more precious. The aforementioned book of ostensibly familiar material, written in a deceptively easy style, was in fact a highly exploratory effort of three authors shortly destined to move to other research areas. Modelling and scaling were at the heart of this boundary-pushing enterprise.

In a wider engineering context, significant basic questions remain. What system model should be chosen? Also, the frequency domain is a greatly neglected area. It is, however, an area of significant interest to model developers in diverse applications where high-frequency and low-frequency models are sought. In Chapter 3 and Chapter 4, the frequency domain is not just breached – it is conquered. All results are new. The basic question of which nonlinear singularly perturbed system model to choose is examined in Chapter 5. New directions are established.

A few remarks on Part 2 of the book are appropriate. Part 2 is aimed squarely at a general engineering audience whose primary interests lie in the modelling of dynamical systems. It is possible that scientists, mathematicians with sharpened pencils, and stiff differential equation experts may also have their appetites whetted. As in Part 1, Part 2 – with its physical narrative and due care of historical context – is conversational in style. It is naturally so in order that engineers, faced with other pressing demands on their time, can readily absorb its contents. This should also be immensely helpful to postgraduate students. Some of the material has separately benefitted from the comments of reviewers.

It is commonly assumed, but rarely stated, that the singularly perturbed dynamical systems under study are physically realisable; that is to say, causal. This is an underlying theme of Part 1. System causality is an explicit requirement throughout Part 2 for all values of singular perturbation parameter epsilon greater than or equal to zero. In the frequency domain, this fundamental assumption of physical realisability requires that the system transfer function be at least proper for all values of epsilon greater than or equal to zero.

THREE

LOW-FREQUENCY AND HIGH-FREQUENCY TRANSFER-FUNCTION SYSTEM MODELS

SUMMARY

This chapter provides new *generic* high-frequency and low-frequency models for transfer-function systems, as they might assist the engineer in their *specific* modelling task. The emphasis is on first principles, with several examples, such that exact high-frequency and low-frequency models are developed in a physical and transparent manner. The low-frequency transfer-function model is defined in the low-frequency s-scale as $G(s, p, \varepsilon)$, where the complex frequency s is 'order of one' and the scaled complex frequency $p = \varepsilon s$ is an 'order of ε' small perturbation. Correspondingly, the high-frequency transfer-function model is defined in the high-frequency p-scale as $G(p, 1/s, \varepsilon)$, where the complex frequency p is 'order of one' and the scaled reciprocal complex frequency $1/s = \varepsilon/p$ is an 'order of ε' small perturbation. A striking advantage of the frequency scaling approach taken is that it naturally extends the operational calculus of Heaviside to two-frequency-scale passive lumped electrical networks and their analogues. Should inductance and capacitance be parasitic, their reactance is conveniently modelled as εL and $1/\varepsilon s C$, respectively. Parasitic resistance εR is also readily treated. Chapter 3 concludes with a detailed discussion and comparison with related work.

3.1 INTRODUCTION

High-frequency and low-frequency models of system dynamical behaviour are common in science and engineering. Their use arises from the desire to view the salient dynamics of the particular dynamical system in a chosen frequency range of interest. The dynamics of the system model in a different frequency range may not be completely neglected, but their role is necessarily a subsidiary one in the chosen frequency range of concern.

As an example, the electrical network specialist may be interested in developing a model of network dynamics in the high-frequency range. While the network dynamics in the low-frequency range may not be completely disregarded, they necessarily play a secondary role in the chosen high-frequency range of interest. Other examples are models of power electronic devices that operate in the high-frequency electromagnetic band, typically anything from a few kilohertz to a few megahertz, whereas the connecting power system operates in the low-frequency electromechanical band, typically 1 Hz. Viewed in the high-frequency electromagnetic band of the power electronic device, the rest of the power system is modelled as a quasi-steady-state resonant *RLC* network (O'Reilly, Wood and Osauskas, 2003).

Finally, control system models are low-pass in nature and thereby operate in the low-frequency range. In that chosen low-frequency range, destabilising unmodelled high-frequency dynamics are attenuated (Bode, 1940; Kokotović, Khalil and O'Reilly, 1999; O'Reilly, 1986).

The purpose of this chapter is to present, in a generic sense, new high-frequency and low-frequency models for general single-input single-output (SISO) transfer-function systems. Aficionados of singular perturbation theory will, of course, recognise concepts and ideas from that field (Kokotović, Khalil and O'Reilly, 1999). The development, however, is simple, direct and self-contained. It should therefore be of primary interest to engineers desirous of generic high-frequency and low-frequency models for their chosen application.

That said, the question arises as to the relationship between a high-frequency range and a low-frequency range, if one is to develop dynamical models suitable for either frequency range. Going further, one might then ask as to the relationships between high-frequency models, low-frequency models and the overall full-frequency-range dynamical model itself. These are profound modelling issues, and their resolution lies in the establishment of a precise relationship between the high-frequency and low-frequency ranges. That relationship is known as the scaling of the two complex frequency ranges, and it is discussed in detail in Kokotović, Khalil and O'Reilly (1999).

The key to this frequency scaling issue actually lies within everyday engineering experience. Consider the dynamical response of a stable linear time-invariant system to a step input. Assuming zero initial conditions, the dynamical response typically takes the form of an initial large fast transient in a 'stretched' time $\tau = t/\varepsilon$, where ε is a small positive scaling parameter, followed by a smaller slow time evolution in time t to a steady-state value.

The frequency response counterpart typically takes the form of an initial gradual low-frequency gain roll-off in the reciprocal complex frequency $1/s$ near and above break frequencies of order one. This is followed by an additional sharper high-frequency gain roll-off in $1/s$ near and above break frequencies of order $1/\varepsilon$. The small scaling parameter ε is the ratio of slow to fast speeds of the system response.

By the scaling theorem, where the time-function $f(t)u(t)$ has the one-sided Laplace transform $F(s)$, that initial $1/\varepsilon$ large fast transient $(1/\varepsilon)f(t/\varepsilon)u(t)$ has the Laplace transform $F(\varepsilon s)$ in the high-frequency scale $p = \varepsilon s$. Expressed equivalently, the 'order of one' fast transient $f(t/\varepsilon)u(t)$ in the time domain has an 'order of ε' small Laplace transform $\varepsilon F(\varepsilon s)$ in the frequency domain. The scaling theorem speaks to a 'duality'[3] between the time domain and frequency domain, which is particularly useful when one considers either two-time-scale or two-frequency-scale systems. In particular, scaling of the complex frequency s by the small positive scaling factor ε in a two-frequency-system is the 'dual' of the scaling of time t by ε in the standard singular perturbation model (Kokotović, Khalil and O'Reilly, 1999).

This frequency scaling allows for the development of exact high-frequency and low-frequency models in a physical and transparent manner. The low-frequency transfer-function model is defined in the low-frequency s-scale as $G(s, p, \varepsilon)$, where the complex frequency s is 'order of one' and the scaled complex frequency $p = \varepsilon s$ is an $O(\varepsilon)$ small perturbation. Also, the high-frequency transfer-function model is defined in the high-frequency p-scale as $G(p, 1/s, \varepsilon)$, where the complex frequency p is 'order of one' and the scaled reciprocal complex frequency $1/s = \varepsilon/p$ is an $O(\varepsilon)$ small perturbation. An advantage of this frequency scaling approach is that it naturally extends the operational calculus of Heaviside to passive lumped electrical networks with parasitic elements, and to their analogues.

Chapter 3 is organised as follows. Section 3.2 defines what is meant by frequency scaling and two-frequency-scale system models in particular.

---

3   A parlour trick for engineers: sketch a typical first-order or second-order step response on a sheet of paper. Turn the sheet of paper over and hold it up to the light. On the reverse, lo and behold, the typical magnitude frequency-response of a stable first-order or second-order system (Professor W. E. Leithead, past private communication).

New high-frequency and low-frequency transfer-function system models are presented in Section 3.3. Section 3.4 presents a new operational calculus of two-frequency-scale passive lumped electrical networks, and their mechanical, fluidic and thermal analogues. Further illustrative examples of high-frequency and low-frequency system models are provided in Section 3.5. Section 3.6 presents a detailed discussion and comparison with related work. Conclusions are outlined in Section 3.7.

3.2   TWO-FREQUENCY-SCALE SYSTEM MODELS

The purpose of this chapter is to develop generic transfer-function models of dynamical systems in the frequency range of interest to the modeller. In order to achieve that, it is first necessary to describe what a two-frequency-scale system model is.

Consider the linear system transfer-function model $G(s, \varepsilon)$ from the single input $u(s)$ to the single output $y(s)$ given by

$$y(s,\varepsilon) = G(s,\varepsilon)u(s) \tag{3.1}$$

where s is the complex frequency, $G(s, \varepsilon)$ is rational in s and ε is a small positive scaling parameter. Both the numerator and denominator polynomials of the system transfer-function model $G(s, \varepsilon)$ in (3.1) are assumed to be of the factored form

$$P(s,\varepsilon) = \prod_{i=1}^{N}(s-s_i) \tag{3.2}$$

where the polynomial roots s_i are either real or occur in complex-conjugate pairs.

Assumption 2.1: The system transfer-function model $G(s, \varepsilon)$ in (3.1) is at least proper for all $\varepsilon \geq 0$ for physical realisability.

As stated, the transfer-function system description $G(s, \varepsilon)$ of (3.1) is quite general. The dependence of $G(s, \varepsilon)$ upon a small parameter ε, at face value, is unremarkable. If, however, one is seeking a two-frequency-scale system model, that small parameter ε needs to be used to scale *some* of the complex frequencies s in the system transfer function $G(s, \varepsilon)$ appropriately. The required scaling is simple. One merely scales the complex frequency s by the scaling parameter ε to obtain the scaled complex frequency $p = \varepsilon s$.

In this way, the polynomial $P(s, \varepsilon)$ of (3.2), written in s-scale, can be rewritten in $p = \varepsilon s$ scale as

$$P(p,\varepsilon) = \left(1/\varepsilon^N\right)\prod_{i=1}^{N}(p - p_i)$$

(3.3)

where the polynomial roots p_i are either real or occur in complex-conjugate pairs.

Thus far, all scaling has been in one frequency scale: the familiar s-scale or the less familiar p-scale. The system transfer function $G(s, \varepsilon)$ or $G(p, \varepsilon)$ is a conventional one-scale model. It is when the small parameter ε is used to scale some, but not all, of the complex frequencies s that we arrive at the following definition of a two-frequency-scale system model.

Definition 2.1 – Two-frequency-scale model: Under Assumption 2.1, the system transfer-function model $G(s, \varepsilon)$ in (3.1) is two-frequency scale if and only if $G(s,\varepsilon) = G(s,p,\varepsilon)$, such that

$$G(s,p,\varepsilon) = G_1(s,\varepsilon)G_2(p,\varepsilon)$$

(3.4)

where s is the complex frequency and $p = \varepsilon s$ is a scaled complex frequency defined by the small positive scaling factor ε. The numerator and denominator polynomials in s of the transfer function $G_1(s,\varepsilon)$ are as in (3.2), whereas the numerator and denominator polynomials in $p = \varepsilon s$ of the transfer function $G_2(p,\varepsilon)$ are as in (3.3).

Remark 2.1: A property of Definition 2.1 is that the poles and zeros of the two-frequency-scale system model $G(s, p, \varepsilon)$ in (3.4) consists of the $O(1)$ poles and zeros of the transfer function $G_1(s,\varepsilon)$, and in s-scale, the $O(1/\varepsilon)$ poles and zeros of the transfer function $G_2(p,\varepsilon)$.

Definition 2.1 of a two-frequency-scale transfer-function system model is new. It is straightforward. It is equivalent to those of Luse and Khalil (1985) and Chaplais and Alaoui (1996), and it is implicit in all previous two-frequency-scale work. Definition 2.1 places scaling first and foremost as a modelling tool if one is to satisfactorily model a transfer-function system in a two-frequency-scale format. This scaling of *some* complex frequencies s by ε to define a two-frequency-scale system is the dual of the scaling of time t by ε to define a singularly perturbed system in standard explicit form (Kokotović, Khalil and O'Reilly, 1999).

Remark 2.2: Henceforth, Definition 2.1 conveniently allows us to describe a two-frequency-scale system transfer function simply as $G(s, p, \varepsilon)$, where s and $p = \varepsilon s$ are the two complex frequencies such that, given the first complex frequency s in s-scale, the second complex variable consists of s scaled by ε to give p. This is similar to multi-scale analysis (Kevorkian and Cole, 1996), where both variables s and p are treated as independent variables, even though they are not since $p = \varepsilon s$.

In Definition 2.1, the scaling of the complex frequency s by the small positive parameter ε plays a very important role. The small scaling parameter ε could represent a physical parameter of the system or it could simply be the ratio of a small system time constant to a large system time constant. Again, the choice of small parameter ε could be at the modeller's discretion in separating the two frequency bands of interest: one in complex frequency s and one in complex frequency $p = \varepsilon s$.

A summary of this two-frequency scaling is depicted in Figure 3.1.

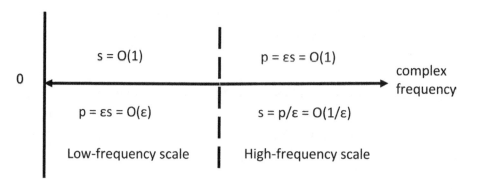

Figure 3.1 The two frequency scales s *and* p

Remark 2.3: A vector function $f(t, \varepsilon) \in R^n$ is said to be $O(\varepsilon)$ over an interval $[t_1, t_2]$ if there exist positive constants k and ε^* such that $\|f(t, \varepsilon)\| \le k\varepsilon$ for all $\varepsilon \in [0, \varepsilon^*]$ and for all $t \in [t_1, t_2]$, where $\|.\|$ denotes the Euclidean norm.

Henceforth, we consider $O(\varepsilon)$ as being an 'order of magnitude relation', valid for 'ε sufficiently small'.

Remark 2.4: In the low-frequency s-scale, the complex frequency s is of 'order of one' or $O(1)$; as such, for ε small, the scaled complex frequency $p = \varepsilon s$ is of 'order of epsilon' or $O(\varepsilon)$ small or low frequency. In the high-frequency $p = \varepsilon s$ scale, the complex frequency p is $O(1)$; as such, for ε small, the complex frequency $s = p/\varepsilon$ is $O(1/\varepsilon)$ large or high frequency. See Figure 3.1.

Remark 2.5: In the low-frequency s-scale, where the complex frequency s is $O(1)$, there exists an upper frequency ω_1 such that $s \in [0, j\omega_1]$. In other words, the interval $[0, j\omega_1]$ defines the band of low frequencies, where the upper frequency ω_1 is specified by the modeller. In the high-frequency p-scale, where the complex frequency p is $O(1)$, there exists a lower frequency ω_2 such that $p \in [j\omega_2, \infty)$. In other words, the interval $[j\omega_2, \infty)$ defines the band of high frequencies where the lower frequency ω_2 is specified by the modeller.

Definition 2.1 and Figure 3.1 have naturally been expressed in terms of the two complex frequencies s and p for the system transfer-function model $G(s, \varepsilon)$. As we shall see in Section 3.3, this is particularly useful for the development of a low-frequency system model.

Alternatively, should a high-frequency system model be of interest, it is particularly useful to express Definition 2.1 in terms of the complex frequency p and the reciprocal complex frequency $1/s$ for the system transfer-function model $G(p, \varepsilon)$. For that, it is necessary to rewrite the polynomial $P(s, \varepsilon)$ of (3.2) in terms of $1/s$ as

$$P\left(1/s,\varepsilon\right)=\left(1/s\right)^{-N}\prod_{i=1}^{N}\left(1/d_i\right)\left(d_i-1/s\right) \tag{3.5}$$

where the polynomial roots d_i are either real or occur in complex-conjugate pairs. The results dual to Definition 2.1 are as follows.

Definition 2.2 – Dual two-frequency-scale model: Under Assumption 2.1, the system transfer-function model $G(p, \varepsilon)$ in (3.1) is two-frequency-scale if and only if $G\left(p,\varepsilon\right)=G\left(p,1/s,\varepsilon\right)$ such that

$$G(p,1/s,\varepsilon)=G_1(p,\varepsilon)G_2(1/s,\varepsilon) \qquad\qquad (3.6)$$

where p is the complex frequency and $1/s = \varepsilon/p$ is a scaled reciprocal complex frequency defined by the small positive scaling factor ε. The numerator and denominator polynomials in p of the transfer function $G_1(p,\varepsilon)$ are as in (3.3), whereas the numerator and denominator polynomials in $1/s = \varepsilon/p$ of the transfer function $G_2(1/s,\varepsilon)$ are as in (3.5).

Remark 2.6: In Definition 2.2, the reciprocal complex frequency $1/s$ is the dual of the complex frequency p in Definition 2.1.

A summary of this two-frequency scaling is depicted in Figure 3.2.

It is observed in Definition 2.2 that, given the first complex frequency p in p-scale, the second complex variable consists of the reciprocal of the complex frequency p scaled by ε to give $1/s$. This is the p-scale dual of Definition 2.1 in s-scale.

Remark 2.7: In the high-frequency p-scale, the complex frequency p is $O(1)$; as such, for ε small, the scaled reciprocal complex frequency $1/s = \varepsilon/p$ is $O(\varepsilon)$ small. In the low-frequency s-scale, the complex frequency p is $O(\varepsilon)$; as such, for ε small, the reciprocal complex frequency $1/s = \varepsilon/p$ is $O(1)$. See Figure 3.2.

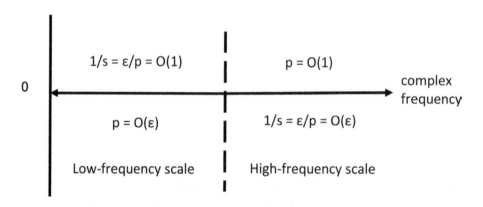

Figure 3.2 *The two frequency scales p and 1/s*

3.3   LOW-FREQUENCY AND HIGH-FREQUENCY SYSTEM MODELS

In this section, we develop appropriate generic transfer-function models of dynamical systems specific to the frequency range of interest to the modeller. That frequency range is either low frequency or high frequency. Let us therefore start with the low-frequency system model first.

Should the modeller wish to develop a system model in the low-frequency scale, it is, of course, necessary to restrict attention to the low-frequency band $s \in [0, j\omega_1]$, where the complex frequency s is $O(1)$, as in Remark 2.5. Then, by Definition 2.1, one immediately has that a low-frequency system model is defined as in the next definition.

Definition 3.1 – Low-frequency model: Under Assumption 2.1, the system transfer-function model $G(s, \varepsilon)$ in (3.1) is a low-frequency model if and only if $G(s,\varepsilon)=G(s,p,\varepsilon)$ such that

$$G(s,p,\varepsilon)=G_1(s,\varepsilon)G_2(p,\varepsilon) \qquad (3.7)$$

and there exists an upper frequency ω_1 such that the complex frequency s is confined to the lower-frequency range $s \in [0, j\omega_1]$ where the scaled complex frequency $p = \varepsilon s$ for some small positive scaling factor ε. The numerator and denominator polynomials in s of the transfer function $G_1(s,\varepsilon)$ are as in (3.2), whereas the numerator and denominator polynomials in $p = \varepsilon s$ of the transfer function $G_2(p,\varepsilon)$ are as in (3.3).

The low-frequency system model $G(s, p, \varepsilon)$ of (3.7) is *exact*. There are no approximations.

In the low-frequency system model $G(s, p, \varepsilon)$ of (3.7), it is observed that the complex frequencies s participate *strongly* as $O(1)$, whereas the complex frequencies $p = \varepsilon s$ participate but *weakly* as $O(\varepsilon)$. This is as it should be in the low-frequency s-scale, where the complex frequency s is $O(1)$ and the complex frequency p is but a *perturbation*. It is consistent with the modeller's desire to focus attention on the low-frequency range of interest, with a passing glance at the subsidiary high-frequency range.

3.3.1 A SIMPLE LOW-FREQUENCY MODEL APPROXIMATION

Suppose one were then to *approximate* this low-frequency model $G(s, p, \varepsilon)$ of (3.7) by setting the complex frequency $p = \varepsilon s$ equal to zero. In this low-frequency s-scale, this is equivalent to approximating the small positive scaling parameter ε by $\varepsilon = 0$, since the complex frequency s is $O(1)$. By Assumption 2.1, the approximate model $G(s, 0, 0)$ of (3.7) is at least proper, and hence physically realisable. In this approximate low-frequency system model $G(s, 0, 0)$, the complex frequencies s still participate strongly as $O(1)$, but the complex frequencies $p = \varepsilon s$ do *not* participate at all.

The original *exact* low-frequency model $G(s, p, \varepsilon)$ of Definition 3.1 is the better low-frequency system model, especially when the scaling parameter ε is small but *not* close to zero. It is recalled that the small positive scaling parameter ε is either known, estimated or at the discretion of the modeller when they decide on appropriate frequency scales.

Let us now consider the dual high-frequency system model. Should the modeller wish to develop a system model in the high-frequency scale, it is of course necessary to restrict attention to the high-frequency band $p \in [j\omega_2, \infty)$, where the complex frequency p is $O(1)$, as in Remark 2.5. Then, by Definition 2.2, one immediately has that a high-frequency system model is defined as in the next definition.

Definition 3.2 – High-frequency model: Under Assumption 2.1, the system transfer-function model $G(p, \varepsilon)$ in (3.1) is a high-frequency model if and only if $G(p,\varepsilon) = G(p,1/s,\varepsilon)$ such that

$$G(p,1/s,\varepsilon) = G_1(p,\varepsilon)G_2(1/s,\varepsilon) \tag{3.8}$$

and there exists a lower frequency ω_2 such that the complex frequency p is confined to the upper-frequency range $p \in [j\omega_2, \infty)$ where the scaled complex frequency $p = \varepsilon s$ for some small positive scaling factor ε. The numerator and denominator polynomials in p of the transfer function $G_1(p,\varepsilon)$ are as in (3.3), whereas the numerator and denominator polynomials in $1/s = \varepsilon/p$ of the transfer function $G_2(1/s,\varepsilon)$ are as in (3.5).

The high-frequency system model $G(p, 1/s, \varepsilon)$ of (3.8) is *exact*. There are no approximations.

In the high-frequency system model $G(p, 1/s, \varepsilon)$ of (3.8), it is observed that the complex frequencies p participate *strongly* as $O(1)$, whereas the reciprocal complex frequencies $1/s = \varepsilon/p$ participate but *weakly* as $O(\varepsilon)$. Put another way, the reciprocal complex frequencies $1/s$ have decayed to within $O(\varepsilon)$. Again, this is as it should be in the high-frequency p-scale, where the complex frequency p is $O(1)$ and the reciprocal complex frequency $1/s$ is but a *perturbation*. It is consistent with the modeller's desire to focus attention on the high-frequency range of interest, with a passing glance at the subsidiary low-frequency range.

3.3.2   A SIMPLE HIGH-FREQUENCY MODEL APPROXIMATION

Suppose one were then to *approximate* this high-frequency model $G(p, 1/s, \varepsilon)$ of (3.8) by setting the reciprocal complex frequency $1/s = \varepsilon/p$ equal to zero. In this high-frequency p-scale, this is equivalent to approximating the small positive scaling parameter ε by $\varepsilon = 0$, since the complex frequency p is $O(1)$. By Assumption 2.1, the approximate model $G(p, 0, 0)$ of (3.8) is at least proper, and hence physically realisable. In this approximate high-frequency system model $G(p, 0, 0)$, the complex frequencies p still participate strongly as $O(1)$, but the reciprocal complex frequencies $1/s = \varepsilon/p$ do *not* participate at all. Again, this approximation of the small positive scaling parameter ε by zero corresponds to an infinite separation of frequency bands.

The original *exact* high-frequency model $G(p, 1/s, \varepsilon)$ of Definition 3.2 is the better high-frequency system model, especially when the scaling parameter ε is small but *not* close to zero. Again, it is recalled that the small positive scaling parameter ε is either known, estimated, or at the discretion of the modeller when they decide on appropriate frequency scales.

Example 3.1: Consider the unity negative-feedback control system of Figure 3.3.

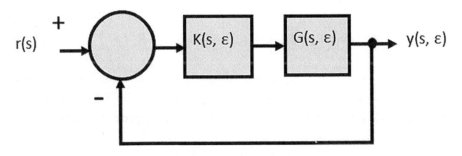

Figure 3.3 A unity negative-feedback control system

A common controller $K(s, \varepsilon)$ in Figure 3.3, which is particularly widespread in the process industries, is the proportional integral derivative (PID) or three-term controller described by

$$K(s,\varepsilon)=K_P + K_D s + K_I / s \qquad (3.9)$$

where the proportional term K_P speeds up the closed-loop system response $y(s, \varepsilon)$, the derivative term $K_D s$ provides damping or stabilising action, and the integral term K_I / s reduces the steady-state error.

Inspection of the three-term controller $K(s, \varepsilon)$ in (3.9) suggests that an analysis in the separate frequency scales of $s = O(1)$ and $p = \varepsilon s = O(1)$ may shed further light on the relative strength of the control action provided by each of the three terms of $K(s, \varepsilon)$.

By Definition 3.1, in the low-frequency $s = O(1)$ scale, the controller $K(s, \varepsilon)$ in (3.9) is rewritten as

$$K(s,\varepsilon)=K(s,p,\varepsilon)=K_P + K_D p + K_I / s \qquad (3.10)$$

where s is $O(1)$, $K_D s$ is recast as $K_D p$ to differentiate it from K_I / s, and $p = \varepsilon s$ is $O(p)$ or $O(\varepsilon)$ small. Consequently, in the low-frequency $s = O(1)$ scale, the derivative action term $K_D p$ is $O(p)$ or $O(\varepsilon)$ weak, and the controller $K(s, \varepsilon)$ in (3.10) can be approximated by

$$K(s,0,0)=K_P + K_I / s \qquad (3.11)$$

At low frequency therefore in $K(s, \varepsilon)$ of (3.10), both the proportional term K_P and the integral term K_I / s are strong, and both reduce the steady-state error.

By Definition 3.2, in the high-frequency $p = \varepsilon s = O(1)$ scale, the controller $K(s, \varepsilon)$ in (3.9) is rewritten as

$$K(s,\varepsilon)=K(p,1/s,\varepsilon)=K_P + K_D p + K_I / s \qquad (3.12)$$

where p is $O(1)$ and $1/s = \varepsilon/p$ is an $O(\varepsilon)$ perturbation. Consequently, in the high-frequency $p = \varepsilon s = O(1)$ scale, the integral action term K_I / s is $O(1/s)$ or $O(\varepsilon)$ weak, and the controller $K(s, \varepsilon)$ in (3.12) can be approximated by

$$K(p,0,0)=K_P + K_D p \qquad (3.13)$$

At high frequency, therefore, in $K(s, \varepsilon)$ of (3.12), both the proportional term K_p and the derivative action term $K_D p$ are strong.

3.4 THE OPERATIONAL CALCULUS OF NETWORKS WITH STRAY OR PARASITIC ELEMENTS

Passive lumped electrical networks are often used as analogues in many branches of science and engineering. Such networks have found particular use in the development of models involving either mechanical motion, heat transfer or fluid flow. What these network models share in common are energy-storage elements that are analogous to the inductors and capacitors of electrical networks. The reactances of these energy-storage elements are highly frequency dependent, and so their dynamic behaviours within networks, at both high and low frequency, are of special interest.

Within the familiar operational calculus of networks, the complex frequency s and the reciprocal complex frequency $1/s$ are treated as differential and integral operators, respectively. Therefore, it is to electrical networks containing inductors L, resistors R and capacitors C that we visit in the light of Definition 2.1 and Definition 2.2 of Section 3.2.

Our attention is first arrested by the fact that the inductive reactance sL and the capacitive reactance $1/sC$ bear an uncanny resemblance to the s and $1/s$ terms in Definition 2.1 and Definition 2.2. This resemblance becomes stronger when the inductance $L = \varepsilon L'$, for $L' = O(1)$, is $O(\varepsilon)$ small, and the capacitance $C = \varepsilon C'$, for $C' = O(1)$, is $O(\varepsilon)$ small. In that case, these *parasitic* or *stray* storage elements L and C have their respective reactances scaled as $\varepsilon s L'$ and $1/\varepsilon s C'$.

Let us first focus on the parasitic inductive storage element $L = \varepsilon L'$ exhibited in s-scale, as in Figure 3.4.

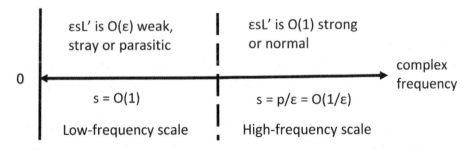

Figure 3.4 Energy-storage inductive reactance $\varepsilon s L'$ in s-scale

And let us now focus on the parasitic capacitive storage element $C = \varepsilon C'$ exhibited in p-scale, as in Figure 3.5.

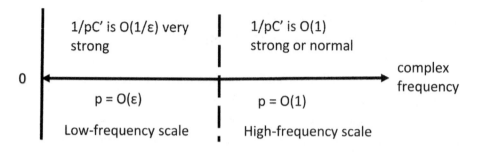

Figure 3.5 Energy-storage capacitive reactance $1/\varepsilon s C'$ in p-scale

Remark 4.1: Simple approximations for the parasitic inductor reactance $\varepsilon s L'$ and the parasitic capacitor reactance $1/\varepsilon s C'$ in the low-frequency $s = O(1)$ scale or $p = O(\varepsilon)$ scale in Figure 3.4 and Figure 3.5 are, respectively, the short circuit and the open circuit obtained by setting $p = \varepsilon s = 0$, or equivalently, by setting the scaling parameter $\varepsilon = 0$.

The presence of storage elements in a network model does not in and of itself suggest, of course, that the network will be two-frequency scale in the sense of Definition 2.1. One signature of when a network is two-frequency scale is when one or more storage elements are parasitic or of $O(\varepsilon)$ relative to the other $O(1)$ storage elements. There may, however, be other tell-tale signs of a two-frequency-scale network. For example, a resistance that is very small relative to other resistances is also a candidate.

By way of illustration, let us therefore recast Example 2.3 and Example 2.4 of Kokotović, Khalil and O'Reilly (1999) in the frequency domain in the next two circuit examples, Example 4.1 and Example 4.2.

Example 4.1: Consider the impedance of the RC circuit in Figure 3.6(a), where R_1, R_2, C_1 and C_2 are all $O(1)$, and ε is a small scaling parameter. In effect, the capacitance εC_2 is parasitic in the low-frequency $s = O(1)$ scale, where $p = \varepsilon s$ is $O(\varepsilon)$.

From Figure 3.6(a), two frequency scales, s and $p = \varepsilon s$, are immediately identified in the sense of Definition 2.1. In the low-frequency $s = O(1)$ scale, Figure 3.6(a) depicts an *exact* low-frequency circuit model in the sense of

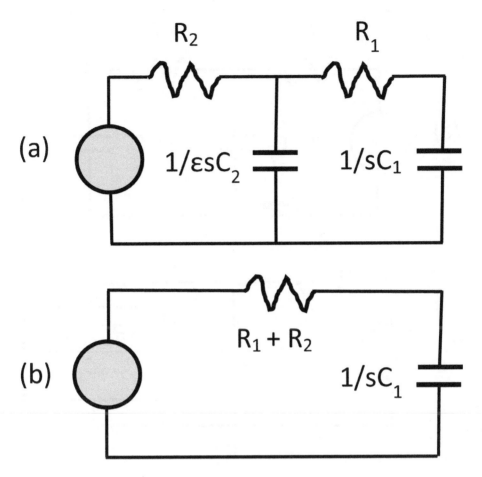

Figure 3.6 *Two-frequency-scale exact circuit model (a) and low-frequency approximate circuit model (b) for parasitic εC_2*

Definition 3.1. The simple low-frequency approximate circuit model in Figure 3.6(b) is easily obtained by replacing the $1/\varepsilon s C_2$ reactance with an open circuit, as in Remark 4.1.

Example 4.2: Consider the impedance of the RC circuit in Figure 3.7(a), where R_1, R_2, C_1 and C_2 are all $O(1)$, and ε is a small scaling parameter. In effect, the resistance εR_1 is parasitic.

Unlike Example 4.1, both storage elements C_1 and C_2 are $O(1)$, and so the presence of two frequency scales is not so apparent. The parasitic resistance εR_1 does, however, look more promising. Should one simply combine εR_1 with the

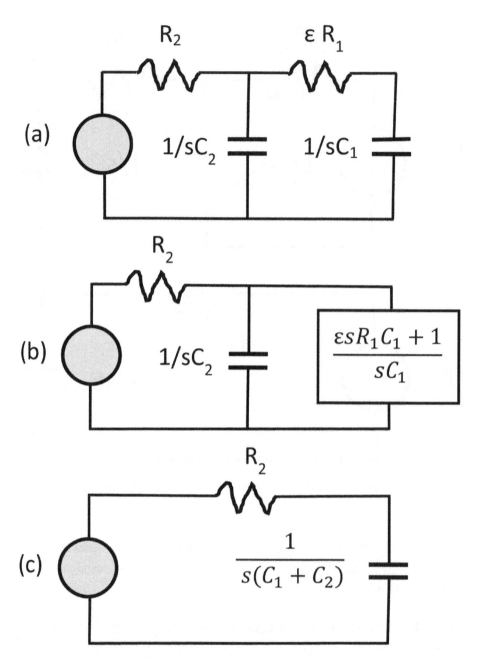

Figure 3.7 Two-frequency-scale exact circuit model (a), equivalent model (b) and low-frequency approximate circuit model (c) for parasitic εR_1

$1/sC_1$ reactance, one arrives at the equivalent circuit model in Figure 3.7(b). This equivalent circuit model is two-frequency-scale in the sense of Definition 2.1.

Also, in the low-frequency $s = O(1)$ scale, Figure 3.7(b) is an *exact* low-frequency circuit model in the sense of Definition 3.1. The simple low-frequency approximate circuit model in Figure 3.7(c) is easily obtained by setting p = εs = 0 in Figure 3.7(b). This is equivalent to directly replacing the parasitic resistance $εR_1$ in Figure 3.7(a) with a short circuit.

Remark 4.2: Connoisseurs of singular perturbation methods will recognise the comparative ease with which the two-frequency-scale circuit model in Figure 3.7(b) of Example 4.2 was achieved. This is because the original circuit model in Figure 3.7(a) is an implicit singularly perturbed system model (Kokotović, Khalil and O'Reilly, 1999). It requires a somewhat tricky change of voltage state variable to render the model in standard explicit singular perturbation form, where ε is the explicit singular perturbation parameter.

Remark 4.3: As an external input-output system description, it is immaterial to the two-frequency-scale transfer-function model as to whether or not it arises from an internal state-space model in explicit or implicit singularly perturbed form.

So much for electrical networks with parasitic elements; in particular, parasitic energy-storage elements. Experts in their chosen fields of mechanical motion, fluid flow or heat transfer will already know the usefulness of electrical network analogues. For the rest of us, the following brief summary of energy-storage elements in Table 3.1 has to suffice.

Table 3.1 Energy-storage-element reactances in electrical, mechanical, fluidic and thermal networks

Electrical network: inductor L and capacitor C	sL	$1/sC$
Mechanical network (translational): mass M and spring constant K	sM	K/s
Mechanical network (rotational): inertia J and torsional spring constant K	sJ	K/s
Fluidic network: inertance I and compliance C_f	sI	$1/sC_f$
Thermal network: thermal capacitance C_t		$1/sC_t$

3.5 LOW-FREQUENCY AND HIGH-FREQUENCY MODEL EXAMPLES

Many dynamical systems in diverse application areas can be modelled effectively by simple combinations of first-order and second-order small-signal subsystem models. Often, these dynamical systems are characterised by the presence of oscillations in different frequency scales. These simple facts are reflected in the following two examples, where the scaling of the complex frequency s by the small positive scaling parameter ε is paramount.

Example 5.1: Consider the second-order system transfer function given by

$$G(s,\varepsilon) = \frac{1}{(1+s)(1+\varepsilon s)} \tag{3.14}$$

By Definition 2.1, the system transfer function (3.14) is two-frequency-scale and is already defined in s-scale.

Consequently, by Definition 3.1, the *exact* low-frequency model for (3.14) is given by

$$G(s,p,\varepsilon) = \frac{1}{(1+s)(1+p)} \tag{3.15}$$

where the modeller specifies an upper frequency ω_1 such that the complex frequency s is $O(1)$ and is confined to the lower-frequency range $s \in [0, j\omega_1]$. The scaled complex frequency $p = \varepsilon s$ is $O(\varepsilon)$ small for small positive scaling factor ε. Consequently, the $p = \varepsilon s$ term in the denominator of the low-frequency model $G(s, p, \varepsilon)$ in (3.15) only participates weakly at $O(\varepsilon)$ in the lower-frequency range $s = O(1)$ or $s \in [0, j\omega_1]$. The low-frequency model $G(s, p, \varepsilon)$ of (3.15) is essentially low-pass in nature.

By Definition 2.2, the system transfer function (3.14) may be redefined in p-scale as

$$G(p,1/s,\varepsilon) = \frac{1/s}{(1+1/s)(1+p)} \tag{3.16}$$

Consequently, by Definition 3.2, the *exact* high-frequency model for (3.14) or (3.16) is given by

$$G(p,1/s,\varepsilon) = \frac{1/s}{(1+1/s)(1+p)} \tag{3.17}$$

where the modeller specifies a lower frequency w_2 such that the complex frequency p is $O(1)$ and is confined to the higher-frequency range $p \in [jw_2, \infty)$. The scaled reciprocal complex frequency $1/s = \varepsilon/p$ is $O(\varepsilon)$ small for small positive scaling factor ε. Consequently, in the high-frequency scale $p = O(1)$ or $p \in [jw_2, \infty)$, the high-frequency model $G(p, 1/s, \varepsilon)$ of (3.17) is only $O(1/s)$ or $O(\varepsilon)$ small.

A simple low-frequency model approximation for the original second-order transfer-function system (3.14) is obtained by setting the complex frequency $p = \varepsilon s$ equal to zero in the low-frequency model $G(s, p, \varepsilon)$ of (3.15) to obtain

$$G(s,0,0) = \frac{1}{(1+s)} \tag{3.18}$$

The transfer function $G(s, 0, 0)$ in (3.18) represents the familiar first-order low-pass filter defined over the frequency range $s \in [0, jw_1]$, where w_1 is the cut-off frequency.

A simple high-frequency model approximation for the original second-order transfer-function system (3.14) or (3.16) is obtained by setting the reciprocal complex frequency $1/s = \varepsilon/p$ equal to zero in the high-frequency model (3.17) to obtain

$$G(p,0,0) = 0 \tag{3.19}$$

over the high-frequency range $p = O(1)$ or $p \in [jw_2, \infty)$. This again shows that the original second-order system transfer function (3.14) is essentially low-pass in nature.

Example 5.2: Consider the third-order system transfer function given by

$$G(s,\varepsilon) = \frac{(s+1)\omega_n^2}{(s+2)\left((\varepsilon s)^2 + 2\xi\omega_n(\varepsilon s) + \omega_n^2\right)} \tag{3.20}$$

for a given damping ratio $\zeta < 1$ and natural frequency w_n.

By Definition 2.1, the system transfer function (3.20) is two-frequency-scale and is already defined in s-scale.

Consequently, by Definition 3.1, *the exact* low-frequency model for (3.20) is given by

$$G\left(s,p,\varepsilon\right) = \frac{\left(s+1\right)\omega_n^2}{\left(s+2\right)\left(p^2 + 2\xi\omega_n p + \omega_n^2\right)} \tag{3.21}$$

where the modeller specifies an upper frequency ω_1 such that the complex frequency s is $O(1)$ and is confined to the lower-frequency range $s \in [0, j\omega_1]$. As such, the scaled complex frequency $p = \varepsilon s$ is $O(\varepsilon)$ small for small positive scaling factor ε. Consequently, the two $p = \varepsilon s$ terms in the denominator of the low-frequency model $G(s, p, \varepsilon)$ in (3.21) only participate weakly at $O(\varepsilon)$ or $O(\varepsilon^2)$ in the lower-frequency range $s = O(1)$ or $s \in [0, j\omega_1]$.

By Definition 2.2, the system transfer function (3.20) may be redefined in p-scale as

$$G\left(p,1/s,\varepsilon\right) = \frac{\left(1+1/s\right)\omega_n^2}{\left(1+2/s\right)\left(p^2 + 2\xi\omega_n p + \omega_n^2\right)} \tag{3.22}$$

Consequently, by Definition 3.2, the *exact* high-frequency model for (3.20) or (3.22) is given by

$$G\left(p,1/s,\varepsilon\right) = \frac{\left(1+1/s\right)\omega_n^2}{\left(1+2/s\right)\left(p^2 + 2\xi\omega_n p + \omega_n^2\right)} \tag{3.23}$$

where the modeller specifies a lower frequency ω_2 such that the complex frequency p is $O(1)$ and is confined to the higher-frequency range $p \in [j\omega_2, \infty)$. As such, the scaled reciprocal complex frequency $1/s = \varepsilon/p$ is $O(\varepsilon)$ small for small positive scaling factor ε.

A simple low-frequency model approximation for the original third-order transfer-function system (3.20) is obtained by setting the complex frequency $p = \varepsilon s$ equal to zero in the exact low-frequency model $G(s, p, \varepsilon)$ of (3.21) to obtain

$$G\left(s,0,0\right) = \frac{\left(s+1\right)}{\left(s+2\right)} \tag{3.24}$$

The model transfer function $G(s, 0, 0)$ in (3.24) is proper, thereby satisfying Assumption 2.1, and is defined over the low-frequency range $s \in [0, j\omega_1]$. This model $G(s, 0, 0)$ in (3.24) is certainly a simple approximation of the exact low-frequency model $G(s, p, \varepsilon)$ of (3.21), which contains a higher-frequency resonance at ω_n/ε in s-scale. Should ε be small but not close to zero, this simple model approximation may prove to be unsatisfactory.

A simple high-frequency model approximation for the original third-order transfer-function system (3.20) or (3.22) is obtained by setting the reciprocal complex frequency $1/s = \varepsilon/p$ equal to zero in the exact high-frequency model (3.23) to obtain

$$G(p,0,0) = \frac{\omega_n^2}{\left(p^2 + 2\xi\omega_n p + \omega_n^2\right)} \tag{3.25}$$

The model transfer function $G(p, 0, 0)$ in (3.25) is strictly proper, thereby satisfying Assumption 2.1, and is defined over the high-frequency range $p = O(1)$ or $p \in [j\omega_2, \infty)$. Should the reciprocal complex frequency $1/s$ or ε be sufficiently small, this will be a satisfactory approximation.

3.6 DISCUSSION AND COMPARISON WITH RELATED WORK

As early as the 1920s, stray capacitance was identified as an oscillation stability issue in high-frequency amplifiers (Friis and Jensen, 1924) and electrical networks (Bode, 1945). While these stray or parasitic elements are most usefully modelled by a two-frequency-scale transfer-function system, the first such model was developed much later by Porter and Shenton (1975) for singularly perturbed systems. Other work followed, but it is to the two-frequency-scale system definition for general transfer-function systems of Luse and Khalil (1985) that we now turn our attention.

Luse and Khalil (1985) start with a definition of pole separation, as in a separation into low-frequency and high-frequency poles, in their definition of a two-frequency-scale system, as one might with an implicit singularly perturbed system model. They then show that this is equivalent to a transfer function $H(s, \varepsilon)$ possessing both s and $p = \varepsilon s$ terms in a particular arrangement in their Theorem 2.2. Essentially, their $H(s, \varepsilon)$ is of a form $H(s, p, \varepsilon)$, as in Definition 2.1 here.

This chapter takes the reverse approach in that it starts with Definition 2.1, as one would with an explicit singularly perturbed system model in standard form. This modelling approach is much closer to that initiated in Kokotović, Khalil and O'Reilly (1999) for state-space systems and further developed in Chapter 4 for transfer-function systems. In particular, the scaling of the complex frequency s is similar to the frequency scaling of the characteristic polynomial in Kokotović, Khalil and O'Reilly (1999).

Historically, as in Biernson (1988), high-frequency transfer-function factors have been expressed as $1 + a/s$ such that they tend to unity as s tends to infinity ($1/s$ tends to zero). The low-pass and high-pass properties of two-frequency-scale systems were first identified in Kokotović, Khalil and O'Reilly (1999) and O'Reilly (1986). The works by Luse (1986, 1988) on multiple time scales, by Oloomi and Sawan (1995) on transfer-function zeros, and by Oloomi and Shafai (2004) on impulse response are more in the spirit of Luse and Khalil (1985).

An important paper by Chaplais and Alaoui (1996) on two-time-scale system identification introduces a low-frequency high-frequency cascade transfer-function model that, suitably interpreted, is equivalent to that of Definition 2.1. Also, it is interesting to note that Chaplais and Alaoui (1996) use low-pass and high-pass prefilters to smooth data in separate frequency scales. Applications of this two-frequency-scale approach to battery model identification and aircraft model identification are respectively presented in Hu and Wang (2015) and in Cavalcanti, Kienitz and Kadirkamanathan (2018). Also, the two-frequency work of Shaker (2009) stresses the importance of retaining system phase information.

3.7 CONCLUSIONS

Phrases such as 'trial and error' and 'ad hoc' are sometimes used to describe modelling and design in engineering. The *actual* process of modelling and design, while outwardly looking as if it might be trial and error, is in fact guided by engineering intuition within a framework of basic engineering principles.

An important part of this modelling process is the scaling or the sizing of the model such that the model fits the purpose for which it was intended. This scaling or sizing can be spatial or temporal. The latter type of scaling – temporal scaling – is of particular importance in dynamical systems. It allows the engineer to decide on the time or frequency scale of main interest, and to develop accordingly the model appropriate to that particular time or frequency scale.

The connection between the two frequency scales – the low-frequency scale and the high-frequency scale – is the small positive scaling parameter ε. This scaling parameter ε is a parameter used to scale the complex frequency, and it can be determined, or at least estimated, in a number of ways. Most obviously, the scaling parameter ε could simply be taken as the physical ratio of two system time constants – one small and one large – or more generally, by a natural separation of low-frequency to high-frequency system dynamics.

The modeller, on the other hand, may simply designate the frequency range of relevance to the intended model to be of one frequency scale, with the remaining frequency range to be the other frequency scale. This needs to be done, of course, while retaining the positive scaling parameter ε as sufficiently small. For instance, in automatic control, all frequencies below the open-loop-system gain crossover frequency might be designated 'low-frequency scale', with those frequencies above the gain crossover frequency designated 'high-frequency scale'. Different choices of the small scaling parameter ε give different relationships between the two frequency ranges.

The purpose of this third chapter is to provide new *generic* high-frequency and low-frequency models for transfer-function systems. This should assist with the modelling process outlined previously, as the engineer might approach it in their *specific* modelling task. A new general frequency-domain approach is thereby established. The emphasis throughout is on first principles, such that the required high-frequency and low-frequency models are developed in a physical and transparent manner. The system transfer-function models are *exact* in their chosen frequency range.

Specifically, the low-frequency transfer-function model is defined in the low-frequency s-scale as $G(s, p, \varepsilon)$, where the complex frequency s is $O(1)$ and the scaled complex frequency $p = \varepsilon s$ is an $O(\varepsilon)$ small perturbation. Correspondingly, the high-frequency transfer-function model is defined in the high-frequency p-scale as $G(p, 1/s, \varepsilon)$, where the complex frequency p is $O(1)$ and the scaled reciprocal complex frequency $1/s = \varepsilon/p$ is an $O(\varepsilon)$ small perturbation.

Another striking advantage of the frequency scaling approach taken is that it naturally extends the operational calculus of Heaviside to two-frequency-scale passive lumped electrical networks and their analogues. The stock-in-trade of such electrical networks is the energy-storage elements of inductance L and capacitance C. Should the inductance L and capacitance C be small or parasitic, their reactance is conveniently modelled by εL and $1/\varepsilon s C$, respectively.

Parasitic resistance εR is also readily treated. The connection to high-frequency and low-frequency transfer-function models is apparent.

The two-frequency-scale analysis of this chapter is conducted for causal or physically realisable linear dynamical systems and networks described by one-sided Laplace transform models. Suppose that the condition of causality is relaxed. It should then be possible to extend the analysis, with its simple scaling, to non-causal systems described by Fourier transform models, as found in communications and signal-processing applications.

An array of system transfer functions, defined over the complex field by the system transfer-function matrix $G(s,\varepsilon) = [G_{ij}(s,\varepsilon)]$, is described on an element-by-element basis. Consequently, all results of this chapter extend to system transfer-function matrices for systems containing several inputs and outputs. In terms of notation, the system transfer function $G(s,\varepsilon)$ is replaced by the system transfer-function matrix $G(s,\varepsilon)$, etc.

The chapter concludes with a detailed discussion and comparison with related work in Section 3.6. Beyond simple ε equals zero approximations, nothing has been said about higher-order transfer-function model approximations – the so-called low-frequency and high-frequency corrected models. That forms the subject of the next chapter.

FOUR

LOW-FREQUENCY AND HIGH-FREQUENCY MODEL APPROXIMATIONS FOR TRANSFER-FUNCTION SYSTEMS

SUMMARY

This chapter presents useful new approximate low-frequency and high-frequency transfer-function system models, which are accurate to any specified order of ε accuracy. These approximate transfer-function models are generic, thereby providing direct assistance to the modeller in their chosen application. The approach also provides asymptotic validity for many common model design approximations where small system constants are neglected. System phase is shown to be pivotal where the neglect of phase in a model should be treated with caution.

4.1   INTRODUCTION

In Chapter 3, both low-frequency and high-frequency models have been developed for transfer-function systems. These models, which depend upon a small positive scaling parameter ε, are exact. No approximations are involved. Nonetheless, it has been shown in Chapter 3 that particularly simple approximations are to be had by setting $\varepsilon = 0$ in the transfer-function system model; for instance, in replacing a parasitic resistance εR by its short-circuit approximation.

In this chapter, we take a closer look at the validity of such approximations. At the same time, useful new approximate low-frequency and high-frequency transfer-function system models, which are accurate to any specified order of

ε accuracy, are presented. Sometimes, these simpler model approximations throw key physical features of the dynamical system into sharper relief than would have been the case in the original more complicated transfer-function system model. Also, these approximate transfer-function models are *generic*. They should therefore assist the model builder in the development of models, each with their own particular physical characteristics, specific to the application of interest.

With regard to the validity of model approximations, intuitively speaking, should the scaling parameter ε be very small or close to zero, the approximation of setting $\varepsilon = 0$ should prove to be very satisfactory; that is to say, accurate enough. Accuracy, of course, is in the eye of the beholder: the modeller. Should the scaling parameter be small but not close to zero, the model approximation obtained upon setting $\varepsilon = 0$ is unlikely to be accurate enough. Recourse will then be needed to higher-order approximations in ε so as to obtain the needed order of accuracy. This is achieved by way of Taylor series expansion in ε in the appropriate frequency scale. The $\varepsilon = 0$ model providing $O(\varepsilon)$ accuracy is known as the 'uncorrected model'. A model that retains at least the first term in ε providing at least $O(\varepsilon^2)$ accuracy is known as a 'corrected model'.

In the context of transfer-function systems in the complex domain, there is an additional physical reason for considering a corrected model, which goes beyond increased accuracy. That physical reason concerns the need to retain the phase of the original transfer-function system in its approximation. Although textbooks on circuits and systems give system gain and system phase equal billing, experienced engineers know better: phase information is more important. While system gain causes amplification or attenuation of the input signal, system phase causes temporal displacement of the input signal acted upon. If you like, the direction of the phasor (vector) is more important than its magnitude.

Consider the single-input single-output (SISO) two-frequency-scale system model of Definition 2.1 of Chapter 3 expressed in the complex frequency $s = j\omega$ as

$$G\left(j\omega, j\varepsilon\omega, \varepsilon\right) = G_1\left(j\omega, \varepsilon\right)G_2\left(j\varepsilon\omega, \varepsilon\right) \qquad (4.1)$$

An examination of the arguments of $G_2(j\varepsilon\omega, \varepsilon)$ in (4.1) reveals that the small scaling parameter ε directly scales the complex frequency $p = j\varepsilon\omega$ and hence the phase contribution of $G_2(j\varepsilon\omega, \varepsilon)$. Should one approximate $G_2(j\varepsilon\omega, \varepsilon)$ by setting

$\varepsilon = 0$ in (4.1), one is, in effect, choosing to neglect that phase contribution in its entirety. The resulting $\varepsilon = 0$ model approximation is the uncorrected model $G_2(0, 0)$, which – as a DC gain or steady-state model – does not depend upon j at all. This may well be an acceptable approximation when ε is close to zero. It is likely to result in an unacceptable loss of phase information, however, should the scaling parameter ε be small but not close to zero. In that case, a corrected system model approximation that retains $j\varepsilon\omega$ information by way of asymptotic expansion is desirable. Such a corrected model will retain the requisite phase information to within $O(\varepsilon^2)$ accuracy.

This is the substance of Chapter 4, which is organised as follows. In Section 4.2, a particular form of Definition 2.1 of Chapter 3 where only the denominator polynomial needs to be factored into low-frequency and high-frequency polynomials is introduced. This provides a system transfer function in two-frequency-scale form. A corrected low-frequency transfer-function model is then presented in Section 4.3, with illustrative examples given in Section 4.4. An important property of the corrected model of Section 4.3 is that it preserves the phase of the original full-order transfer-function model. A dual corrected high-frequency transfer-function model is presented in Section 4.5, with illustrative examples. Conclusions are outlined in Section 4.6.

4.2   TWO-FREQUENCY-SCALE LINEAR SISO SYSTEMS

Consider, as in Chapter 3, the strictly proper linear transfer function $G(s, \varepsilon)$ from the single input $u(s)$ to the single output $y(s, \varepsilon)$ given by

$$y(s,\varepsilon) = G(s,\varepsilon)u(s) \tag{4.2}$$

where s is the complex frequency and ε is a small positive parameter. It is assumed that $G(s, \varepsilon)$ is rational in s, and that $G(s, 0)$ is defined and proper. In this way, the transfer function $G(s, \varepsilon)$ in (4.1) assumes the conventional form

$$G(s,\varepsilon) = \frac{M(s,\varepsilon)}{N(s,\varepsilon)} \tag{4.3}$$

where the order of the numerator polynomial $M(s, \varepsilon)$ is less than the order of the denominator polynomial $N(s, \varepsilon)$.

In other words, under the conditions of Definition 2.1 of Chapter 3, we seek a two-frequency-scale system model described by (4.2) and (4.3), where the transfer-function $G(s, \varepsilon)$ assumes the factored form

$$G(s,\varepsilon) = \frac{M(s,\varepsilon)}{N_1(s,\varepsilon)N_2(\varepsilon s,\varepsilon)} \tag{4.4}$$

in which the order of the polynomials $M(s, \varepsilon)$, $N_1(s, \varepsilon)$ and $N_2(\varepsilon s, \varepsilon)$ are respectively r, n and m, with $r < n + m$. It is assumed that the polynomial $N_2(\varepsilon s, \varepsilon)$ is normalised such that it is unity in the limit $p = \varepsilon s = 0$.

The two-frequency-scale model (4.4) is a special case of Definition 2.1 of Chapter 3 where for our purposes only the denominator polynomial needs to be in factored form.

4.3 A CORRECTED LOW-FREQUENCY MODEL AND PRESERVATION OF PHASE

The two-frequency-scale transfer-function system (4.4) is of order $n + m$, where n and m are respectively the orders of the low-frequency denominator polynomial $N_1(s, \varepsilon)$ and the high-frequency denominator polynomial $N_2(\varepsilon s, \varepsilon)$. Our interest in this section is in the development of an appropriate low-frequency approximate model of order n. In accordance with Remark 2.4 of Chapter 3, the low-frequency scale used is where $s = O(1)$ and $p = \varepsilon s$ is $O(\varepsilon)$. The desired reduction in system order from $n + m$ to n is achieved by way of a simple geometric power series expansion in $p = \varepsilon s$ order of ε small.

Recall that, in Section 4.2, the high-frequency transfer function $N_2^{-1}(\varepsilon s, \varepsilon)$ of (4.4) is assumed to be in normalised form. Therefore, we may rewrite (4.4) as

$$G(s,\varepsilon) = G_1(s,\varepsilon)N_2^{-1}(\varepsilon s,\varepsilon) \tag{4.5}$$

where

$$G_1(s,\varepsilon) = M(s,\varepsilon)N_1^{-1}(s,\varepsilon) \tag{4.6}$$

and

$$N_2(\varepsilon s,\varepsilon) = 1 + h(\varepsilon s,\varepsilon) \tag{4.7}$$

where $h(\varepsilon s,\ \varepsilon)$ is a polynomial of order m. The polynomial $h(\varepsilon s,\ \varepsilon)$ in (4.7) is zero in the limit $p = \varepsilon s = 0$.

In equation (4.5), it is recalled from (4.4) of Section 4.2 that the denominator polynomial $N_1(s,\ \varepsilon)$ of $G_1(s,\ \varepsilon)$ is of order n, and the denominator polynomial $N_2(\varepsilon s,\ \varepsilon)$ is of order m. The order of the denominator polynomial of the original full-order transfer function $G(s,\ \varepsilon)$ in (4.5), as in (4.4), is $n + m$.

The following lemma is a restatement of the geometric power series expansion $(1 + h(p,\ \varepsilon))^{-1}$ for $p = \varepsilon s$ sufficiently small.

Lemma 3.1: There exists a positive fixed constant ω_1 such that the higher-frequency polynomial $N_2^{-1}(\varepsilon s,\ \varepsilon)$ is given by

$$N_2^{-1}(\varepsilon s,\varepsilon) = \left(1+h(p,\ \varepsilon)\right)^{-1} = 1 - h(p,\varepsilon) + \ldots + \left(-h(p,\varepsilon)\right)^{k} + O\left(p^{k+1}\right)$$

(4.8)

where for convergence

$$\left|h(p,\varepsilon)\right| < 1 \qquad (4.9)$$

for all $p = \varepsilon s \in [0,\ j\varepsilon\omega_1)$.

Direct application of Lemma 3.1 to the original two-frequency-scale full-order transfer function (4.5) to (4.7) immediately provides for a low-frequency approximation of this full-order transfer function to any order of accuracy $O(p^{k+1})$ for $p = \varepsilon s$ sufficiently small. In other words, we have the following result.

Result 3.1: There exists a positive fixed constant ω_1 such that the two-frequency-scale transfer function $G(s,\ \varepsilon)$ in (4.5) to (4.7) is approximated in the low-frequency scale $s = 0(1)$ according to

$$G(s,\varepsilon) = G_1(s,\varepsilon)\left[1 - h(p,\varepsilon) + \ldots + \left(-h(p,\varepsilon)\right)^{k}\right] + O\left(p^{k+1}\right) \qquad (4.10)$$

where for convergence

$$\left| h\left(p,\varepsilon \right) \right| < 1 \tag{4.11}$$

for all $p = \varepsilon s \in [0, j\varepsilon w_1)$.

Remark 3.1: It is recalled from Remark 2.4 and Figure 3.1 of Chapter 3 that, in the low-frequency scale, s is $O(1)$. Therefore, in $p = \varepsilon s$ scale, all low-frequency $O(p^k)$ approximations in this chapter, including those of Lemma 3.1 and Result 3.1, are in fact $O(\varepsilon^k)$.

Remark 3.2: It is observed in (4.10) of Result 3.1 that the high-frequency denominator polynomial $N_2(\varepsilon s, \varepsilon)$ of (4.5) of order m has disappeared. Therefore, compared to the original transfer-function model $G(s, \varepsilon)$ in (4.5), the order of the denominator polynomial in (4.10) has collapsed from $n + m$ to n. This is analogous to a singular perturbation order collapse from $n + m$ to n in the time domain, where n and m respectively are the orders of the slow and fast subsystems (Kokotović, Khalil and O'Reilly, 1999). The key to this collapse in system order from $n + m$ to n in the frequency domain is the geometric expansion of Lemma 3.1, where the disappearance of $N_2(\varepsilon s, \varepsilon)$ in the transfer-function denominator of (4.5) is compensated for by the introduction of a regular asymptotic expansion in the transfer-function numerator in (4.10) for $p = \varepsilon s$ sufficiently small.

This phenomenon is consistent with the singular perturbation approach in the time domain, where the fast subsystem is initially neglected to lower the system order from $n + m$ to n. Then, after the regularisation of the original system in separate time scales, the effect of the fast subsystem is reintroduced, as it affects the slow subsystem, by way of asymptotic expansion to any order of accuracy in ε.

Remark 3.3: The reduction in system order in Result 3.1 – or more precisely, the reduction in pole-zero excess – comes at a price. The price is that the model approximation may be non-proper. Properness of the system model approximation is, however, required for physical realisability.

Should we retain only the first two terms of the regular asymptotic expansion in (4.10) of Result 3.1, we have the following important result.

Result 3.2: There exists a positive fixed constant ω_1 such that the two-frequency-scale transfer function $G(s, \varepsilon)$ in (4.5) to (4.7) is approximated in the low-frequency scale $s = 0(1)$ according to

$$G(s,\varepsilon)=G_1(s,\varepsilon)\left[1-h(p,\varepsilon)\right]+O\left(p^2\right) \tag{4.12}$$

where for convergence

$$\left|h(p,\varepsilon)\right|<1 \tag{4.13}$$

for all $p = \varepsilon s \in [0, j\varepsilon\omega_1)$.

As before in Result 3.1, it is observed in Result 3.2 that, compared to the original transfer function $G(s, \varepsilon)$ in (4.5), the order of the denominator transfer function in (4.12) has collapsed from $n + m$ to n.

In Result 3.2, the transfer function $G_c(s, \varepsilon)$ in the low-frequency scale $s = O(1)$ given by

$$G_c(s,\varepsilon)=G_1(s,\varepsilon)\left[1-h(p,\varepsilon)\right] \tag{4.14}$$

is known as the first-order *corrected* low-frequency model in that it provides an $O(p^2)$ approximation, or by Remark 3.1 an $O(\varepsilon^2)$ approximation to the original two-frequency-scale transfer function $G(s, \varepsilon)$ in (4.5) to (4.7) for $p = \varepsilon s$ sufficiently small.

Remark 3.4: Since $h(p, \varepsilon)$ in (4.12) and (4.14) will, in general, contain powers of p greater than one, these higher-order terms in p can be discarded so as to ensure properness of the corrected low-frequency model $G_c(s, \varepsilon)$ in (4.14).

By contrast, should we simply set $\varepsilon = 0$ in the original two-frequency-scale full-order transfer-function model in (4.5) to (4.7), we end up with the usual zero-order *uncorrected* low-frequency model

$$G_0(s,0)=G(s,0) \tag{4.15}$$

This uncorrected model $G_0(s, 0)$ is proper by assumption based on (4.2). It provides an $O(p)$ or $O(\varepsilon)$ approximation to the original cascade transfer function $G(s, \varepsilon)$ in (4.5) to (4.7) at low frequency.

Remark 3.5: Since the small positive parameter ε is at best estimated and the coefficients in the transfer-function model $G(s, \varepsilon)$ in (4.5) to (4.7) are themselves inevitably subject to uncertainty, little is to be gained in pursuing corrections to the uncorrected model of order higher than one, as provided by Result 3.2.

An important property of the corrected low-frequency transfer function $G_c(s, \varepsilon)$ in (4.14), according to Result 3.2, is that it *preserves the higher-frequency phase* to within $O(p^2)$ accuracy (as well as the gain) of the original two-frequency-scale full-order transfer-function model $G(s, \varepsilon)$ in (4.5) to (4.7). This is in contradistinction to the uncorrected low-frequency transfer-function model $G_0(s, 0)$ in (4.15), where only the normalised DC-gain value of the high-frequency transfer-function part is retained.

The preservation of higher-frequency phase by the corrected low-frequency transfer-function model $G_c(s, \varepsilon)$ in (4.14) is very important in the frequency response analysis of circuits and systems, not to mention other engineering dynamical systems. This is because the phase of the unmodelled or neglected high-frequency dynamics can be significant, even at lower frequencies that include the open-loop-system gain crossover frequency. Hence, phase assumes a pivotal role in the system dynamics around this gain-crossover frequency region, and it needs to be accounted for. The corrected low-frequency transfer-function model $G_c(s, \varepsilon)$ in (4.14) accomplishes this, unlike the uncorrected *($\varepsilon = 0$)* model $G_0(s, 0)$ of (4.15).

4.3.1 LOW-FREQUENCY MODELLING

Modelling, as emphasised by Kokotović, Khalil and O'Reilly (1999), is at the heart of the analysis and design of dynamical systems. As such, modelling practice assumes many distinct but related forms. One particular form is least-square-error system model identification from measured system data (Söderström and Stoica, 1989; Young, 2011). In a presentation of river flow data, the author noticed that the identified low-order transfer-function model introduced a right-half-plane (RHP) zero at the same frequency as a high-frequency left-half-plane (LHP) pole in the identified high-order transfer-function model. Indeed, a common model-order reduction procedure in industrial servo design practice has been to directly replace high frequency LHP poles by RHP zeros at the same frequency. More generally, frequency-response analysis supported by Result 3.2 allows for the following assertion.

Assertion 3.1: All low-order identified transfer-function models of a dynamical system, if valid, should preserve the phase of higher-order identified transfer-function models of the same system.

In fact, Assertion 3.1 can be extended to *all* low-order transfer-function system models, irrespective of the particular method by which such models have been developed. This is stated in Assertion 3.2.

Assertion 3.2: All low-order transfer-function models of a dynamical system, if valid, should preserve the phase of the higher-order transfer-function models of the same system.

Let us now review all three system models in the light of Result 3.2 and Assertion 3.2: the original model $G(s, \varepsilon)$ of (4.4) or (4.5), the corrected model $G_c(s, \varepsilon)$ of (4.14), and the uncorrected model $G_0(s, 0)$ of (4.15). The choice of appropriate model crucially depends upon the value of the frequency-scaling parameter ε. There are three cases to consider.

Case 3.1: – ε *is very small or close to zero* – wide separation of frequency scales. In $p = \varepsilon s$ scale, the original high-frequency transfer function $N_2^{-1}(\varepsilon s, \varepsilon)$ of (4.5) is practically steady state. The uncorrected model $G_0(s, 0)$ of (4.15) providing an $O(p)$ or $O(\varepsilon)$ approximation to the original system transfer function $G(s, \varepsilon)$ suffices.

Case 3.2: – ε *is small but not close to zero* – reasonable separation of frequency scales. In $p = \varepsilon s$ scale, the original high-frequency transfer function $N_2^{-1}(\varepsilon s, \varepsilon)$ of (4.5) is quasi-steady-state. The corrected model $G_c(s, \varepsilon)$ of (4.14) providing an $O(p^2)$ or $O(\varepsilon^2)$ approximation to the original system transfer function $G(s, \varepsilon)$ suffices.

Case 3.3: – ε *is not particularly small* – poor separation of frequency scales. The approximation of *Result 3.2* breaks down. Use the original system transfer function $G(s, \varepsilon)$ of (4.4) or (4.5).

4.4   LOW-FREQUENCY MODEL EXAMPLES

Example 4.1: This example is taken from Kokotović, Khalil and O'Reilly (1999). Consider the two-frequency-scale system transfer function given by (4.4) or (4.5) as

$$G(s,\varepsilon) = \frac{1}{(s+1)(\varepsilon s+1)} \tag{4.16}$$

Note that the second-order system of (4.16) has a low-frequency pole at $s = -1$, and a high-frequency pole at $s = -1/\varepsilon$ or $p = \varepsilon s = -1$.

From Result 3.2 of Section 4.3, the corrected low-frequency model $G_c(s, \varepsilon)$ is first-order and is given by (4.14) as

$$G_c(s,\varepsilon) = \frac{1-\varepsilon s}{s+1} \tag{4.17}$$

where for convergence[4]

$$|\varepsilon s| < 1 \tag{4.18}$$

This corrected model $G_c(s, \varepsilon)$ provides an $O(p^2)$ or $O(\varepsilon^2)$ approximation to the original second-order transfer function $G(s, \varepsilon)$ of (4.16) for $p = \varepsilon s$ sufficiently small.

The uncorrected low-frequency model $G_0(s, 0)$ in (4.15) is obtained by simply setting $\varepsilon = 0$ in (4.16) to give

$$G_0(s,0) = \frac{1}{s+1} \tag{4.19}$$

and provides an $O(p)$ or $O(\varepsilon)$ approximation to the original transfer function $G(s, \varepsilon)$ of (16).

Comparison of the corrected low-frequency transfer function $G_c(s, \varepsilon)$ in (4.17) with the original transfer function $G(s, \varepsilon)$ in (4.16) is revealing. It shows that the loss of the high-frequency pole $p = \varepsilon s = -1$ through order reduction is compensated

---

4   The suggestion of Professor W. E. Leithead to provide bounds for convergence, rather than just rely on 'for ε sufficiently small' arguments, is gratefully acknowledged.

for by the introduction of the high-frequency zero $p = \varepsilon s = 1$ at the same frequency in the corrected model $G_c(s, \varepsilon)$ of (4.17). Phase is thereby preserved exactly.

A slightly different corrected low-frequency transfer-function model to that of (4.17) is presented by Kokotović, Khalil and O'Reilly (1999) and is given by

$$G_a\left(s,\varepsilon\right) = \frac{1 - \varepsilon s / \left(1 - \varepsilon\right)}{s + 1} \qquad (4.20)$$

The corrected model $G_a(s, \varepsilon)$ of (4.20) is, however, identical to the corrected model $G_c(s, \varepsilon)$ of (4.17) to within $O(p^2)$ or $O(\varepsilon^2)$.

Example 4.2: Consider the two-frequency-scale transfer function of a third-order system given by (4.4) or (4.5) as

$$G\left(s,\varepsilon\right) = \frac{\omega_n^2 \left(\varepsilon s + 1\right)}{\left(s + 1\right)\left(\omega_n^2 + 2\xi\omega_n\left(\varepsilon s\right) + \left(\varepsilon s\right)^2\right)} \qquad (4.21)$$

It is noted that the third-order system of (4.21) has a low-frequency pole at $s = -1$, a high-frequency zero at $s = -1/\varepsilon$, and an under-damped (oscillatory) high-frequency pole pair at

$$s = [-\xi\omega_n \pm j\omega_n(1 - \xi^2)^{\frac{1}{2}}] / \varepsilon \qquad (4.22)$$

for a given damping ratio $\xi < 1$ and natural frequency ω_n.

From Result 3.2 of Section 4.3, a corrected low-frequency model $G_c(s, \varepsilon)$ is first-order and is given by (4.14) as

$$G_c\left(s,\varepsilon\right) = \frac{\left(\omega_n^2 - 2\xi\omega_n\left(\varepsilon s\right) - \left(\varepsilon s\right)^2\right)\left(\varepsilon s + 1\right)}{\left(s + 1\right)\omega_n^2} \qquad (4.23)$$

Neglecting $(\varepsilon s)^2$ terms and above in the numerator of (4.23) in accordance with Remark 3.4 of Section 4.3, we arrive at the corrected low-frequency model $G_c(s, \varepsilon)$ of first order which is proper and is given by

$$G_c\left(s,\varepsilon\right) = \frac{\omega_n^2 - 2\xi\omega_n\left(\varepsilon s\right) + \omega_n^2\left(\varepsilon s\right)}{\left(s + 1\right)\omega_n^2} \qquad (4.24)$$

where for convergence

$$\left| \varepsilon s / \left(\left(\left(1+\xi^2\right)^{\frac{1}{2}} - \xi \right) \omega_n \right) \right| < 1 \qquad (4.25)$$

This corrected model $G_c(s, \varepsilon)$ of (4.24) provides an $O(p^2)$ or $O(\varepsilon^2)$ approximation to the original third-order transfer function $G(s, \varepsilon)$ of (4.21) for $p = \varepsilon s$ sufficiently small.

The uncorrected low-frequency model $G_0(s, 0)$ in (4.15), obtained by setting $\varepsilon = 0$ in $G(s, \varepsilon)$ of (4.21), is given by

$$G_0\left(s,0\right) = \frac{1}{s+1} \qquad (4.26)$$

It is worthwhile examining all three models: the original system third-order model (4.21), the corrected model (4.24), and the uncorrected model (4.26). Inspection of the original third-order model $G(s, \varepsilon)$ of (4.21) reveals a system that has significant dynamics, a second-order system resonance in fact, in the higher-frequency range. Whether or not the use of the lower-order corrected model $G_c(s, \varepsilon)$ of (4.24) will provide a good enough approximation crucially depends on the scaling parameter ε being sufficiently small.

Should in fact the scaling parameter ε be sufficiently small, the resonance will occur at a frequency ω_n/ε in s-scale rather higher than the open-loop system gain crossover frequency. Then the corrected model $G_c(s, \varepsilon)$ of (4.24) will provide the necessary $O(p^2)$ approximation, most notably with respect to higher-frequency phase. Finally, the uncorrected model $G_0(s, 0)$ of (4.26), which does not preserve higher-frequency phase, is likely to be too simple in this regard.

4.5 A CORRECTED HIGH-FREQUENCY MODEL WITH PHASE PRESERVATION

In Sections 4.3 and 4.4, the focus has been on the development of corrected low-frequency system models in which the high-frequency phase is preserved as it affects the lower frequencies of interest. This is highly appropriate, for instance, in the areas of modelling and identification of dynamical systems where a low-order 'dominant' model is sought. However, there are areas – electrical networks, for instance – where high-frequency models are of interest. A study of high-frequency models also provides completeness.

Accordingly, this section develops a high-frequency approximate model for the original two-frequency-scale transfer function (4.4), rewritten in the high-frequency scale $p = \varepsilon s$ where p is $O(1)$, as in Remark 2.7 of Chapter 3. This is in contradistinction to the low-frequency scale of earlier sections where we dealt with regular asymptotic expansions in p where p is $O(\varepsilon)$.

Consistent with Definition 2.2 of Chapter 3, consider the original two-frequency-scale transfer function (4.4) rewritten in the high-frequency scale $p = \varepsilon s$ where p is $O(1)$ as

$$G(p/\varepsilon,\varepsilon) = \frac{M(p/\varepsilon,\varepsilon)}{N_1(p/\varepsilon,\varepsilon)N_2(p,\varepsilon)} \tag{4.27}$$

As in Section 4.2, it is assumed that $G(p / \varepsilon, \varepsilon)$ is strictly proper, and is defined and proper at $\varepsilon = 0$. What is sought is a corrected high-frequency model that approximates $G(p / \varepsilon, \varepsilon)$ of (4.27) in the high-frequency scale $s = p/\varepsilon$ where p is $O(1)$ and s is $O(1/ \varepsilon)$. This is achieved by way of a suitable power series expansion in $1/s$ or ε/p small.

This $1/s$ or ε/p small power series expansion in the high-frequency scale $p = O(1)$ is the *dual* of the $p = \varepsilon s$ small power series expansion in the low-frequency scale $s = O(1)$ of Section 4.3. It proceeds as follows.

Suppose in (4.27) that the n-th order lower-frequency polynomial $N_1(p/\varepsilon, \varepsilon) = N_1(p, \varepsilon)$, exhibited in the high-frequency scale $p = O(1)$, is given by

$$N_1(p) = \left((p/\varepsilon)^n + b_{n-1}(p/\varepsilon)^{n-1} + \cdots + b_0 \right)$$
$$= (p/\varepsilon)^n \left(1 + b_{n-1}(\varepsilon/p) + \cdots + b_0(\varepsilon/p)^n \right)$$
$$= (p/\varepsilon)^n \left(1 + \ell(\varepsilon/p,\varepsilon) \right) \tag{4.28}$$

so that $\ell(\varepsilon/p,\varepsilon)$, as defined in (4.28), is a polynomial of order n in ε/p small that tends to zero as ε tends to zero.

From (4.27) and (4.28), by way of a power series expansion in ε/p small, the lower-frequency polynomial $N_1^{-1}(p, \varepsilon)$, in high-frequency scale $p = O(1)$, is given by

$$N_1^{-1}(p,\varepsilon) = (\varepsilon/p)^n \left(1 - \ell(\varepsilon/p,\varepsilon) + \ldots + \left(-\ell(\varepsilon/p,\varepsilon)\right)^k + O(\varepsilon/p)^{k+1}\right)$$

(4.29)

where for convergence

$$\left|\ell(\varepsilon/p,\varepsilon)\right| < 1$$

(4.30)

Define also in (4.27)

$$G_2(p,\varepsilon) = M(p/\varepsilon,\varepsilon) N_2^{-1}(p,\varepsilon)$$

(4.31)

Therefore, setting $k = 1$ in (4.29), we have in analogy with the low-frequency Result 3.2 of Section 4.3 the following high-frequency result.

Result 5.1: There exists a positive fixed constant ε^* such that the two-frequency-scale transfer function $G(p/\varepsilon, \varepsilon)$ in (4.27) and (4.31) is approximated in high-frequency scale $p = O(1)$ according to

$$G(p/\varepsilon,\varepsilon) = (\varepsilon/p)^n \left(1 - \ell(\varepsilon/p,\varepsilon) + O\left((\varepsilon/p)^2\right)\right) G_2(p,\varepsilon)$$

(4.32)

where for convergence

$$\left|\ell(\varepsilon/p,\varepsilon)\right| < 1$$

(4.33)

for all $\varepsilon \in (0, \varepsilon^*)$.

Remark 5.1: It is recalled from Remark 2.7 of Chapter 3 that in the high-frequency scale, p is $O(1)$. Therefore, in $\varepsilon/p = 1/s$ scale, all high-frequency $O((\varepsilon/p)^i)$ approximations, including those of (4.29) and Result 5.1, are in fact $O(\varepsilon^i)$.

Remark 5.2: In any particular example, as follows, the overall order of approximation in ε/p in (4.32) can be specified. However, the order of approximation cannot be specified in general since there may be internal cancellations of ε/p terms with p/ε terms from the numerator of $G_2(p,\varepsilon)$ in (4.31).

In Result 5.1, the transfer function $G_h(p,\varepsilon)$ in high-frequency scale $p = O(1)$ given by

$$G_h(p,\varepsilon)=(\varepsilon/p)^n\left(1-\ell(\varepsilon/p,\varepsilon)\right)G_2(p,\varepsilon) \tag{4.34}$$

is the first-order *corrected* high-frequency model in that it provides a corrected order of approximation to the original two-frequency-scale cascade transfer function G(p/ε, ε) in (4.27) for ε/p sufficiently small. It is pointed out however that, unlike the corrected low-frequency model of Section 4.3, the above corrected high-frequency model may not be of lower order than that of the original system model.

By contrast with the corrected high-frequency model of Result 5.1, should we simply set ε=0 in the original two-frequency-scale transfer function (4.27), we end up with the zero-order *uncorrected* high-frequency model in p = O(1) scale

$$G_h(p,0)=G(p/\varepsilon,\varepsilon)\big|_{\varepsilon=0} \tag{4.35}$$

Example 5.1: Consider the two-frequency-scale system transfer function given by

$$G(s,\varepsilon)=\frac{(s+2)}{(s+1)(\varepsilon s+1)} \tag{4.36}$$

rewritten in p = εs scale in ε/p small as

$$G(p/\varepsilon,\varepsilon)=\frac{(p+2\varepsilon)}{(p+\varepsilon)(p+1)}=\frac{(1+2\varepsilon/p)}{(1+\varepsilon/p)(p+1)} \tag{4.37}$$

As in Result 5.1, we have by way of power series expansion in ε/p small that

$$G(p/\varepsilon,\varepsilon) = \frac{\left(1+2\varepsilon/p\right)\left(1-\varepsilon/p+\left(\varepsilon/p\right)^2 -\right)}{\left(p+1\right)} \tag{4.38}$$

where for convergence

$$\left|\varepsilon/p\right| < 1 \tag{4.39}$$

On simplifying equation (4.38), one has

$$G(p/\varepsilon,\varepsilon) = \frac{p+\varepsilon}{p(p+1)} + O\left(\left(\varepsilon/p\right)^2\right) \tag{4.40}$$

The corrected high-frequency model $G_h(p,\varepsilon)$ in $p = O(1)$ scale is thereby given by

$$\dot{G}_h(p,\varepsilon) = \frac{p+\varepsilon}{p(p+1)} \tag{4.41}$$

Alternatively, in $s = p/\varepsilon$ scale, the corrected model $G_h(s,\varepsilon)$ is given by

$$G_h(s,\varepsilon) = \frac{(s+1)}{s(\varepsilon s+1)} \tag{4.42}$$

The uncorrected high-frequency model $G_h(p,0) = G(p/\varepsilon,\varepsilon)\big|_{\varepsilon=0}$ in $p = O(1)$ scale is given by

$$G_h(p,0) = \frac{1}{(p+1)} \tag{4.43}$$

or in $s = p/\varepsilon$ scale by

$$G_h(s,0) = \frac{1}{(\varepsilon s + 1)} \qquad (4.44)$$

Example 5.2: Consider the two-frequency-scale system transfer function given by

$$G(s,\varepsilon) = \frac{(s+1)\omega_n^2}{\left(\omega_n^2 + 2\xi\omega_n s + s^2\right)(\varepsilon s + 1)} \qquad (4.45)$$

rewritten in $p = \varepsilon s$ scale in ε/p small as

$$G(p/\varepsilon,\varepsilon) = \frac{(\varepsilon/p)(1+(\varepsilon/p))\omega_n^2}{(1+2\xi\omega_n(\varepsilon/p)+\omega_n^2(\varepsilon/p)^2)(p+1)} \qquad (4.46)$$

As in Result 5.1, we have by way of power series expansion in ε/p small that

$$G(p/\varepsilon,\varepsilon) = \frac{(\varepsilon/p)\omega_n^2}{(p+1)} + O\left((\varepsilon/p)^2\right) \qquad (4.47)$$

where for convergence

$$\left|2\xi\omega_n(\varepsilon/p)+\omega_n^2(\varepsilon/p)^2\right| < 1 \qquad (4.48)$$

The corrected high-frequency model $G_h(p,\varepsilon)$ in $p = O(1)$ scale is thereby given by

$$G_h(p,\varepsilon) = \frac{\varepsilon\omega_n^2}{p(p+1)} \qquad (4.49)$$

or in $s = p/\varepsilon$ scale by

$$G_h(s,\varepsilon) = \frac{\omega_n^2}{s(\varepsilon s+1)} \qquad (4.50)$$

From (4.46), the uncorrected high-frequency model $G_h(p,0) = G(p/\varepsilon,\varepsilon)\big|_{\varepsilon=0}$

is zero. Essentially, the original system transfer-function $G(s, \varepsilon)$ of (4.45), in high-frequency scale with $\varepsilon/p = 1/s$ small, has rolled off to zero.

4.6 CONCLUSIONS

This chapter presents useful new low-frequency and high-frequency model approximations for transfer-function systems. These transfer-function model approximations are accurate to any specified order of ε accuracy, where ε is a small positive scaling parameter. The approximate transfer-function models are also generic, thereby providing direct assistance to the modeller in their chosen application. In addition, the approach provides asymptotic validity for many common model design approximations where small system constants are neglected.

System phase is shown to be pivotal where the neglect of phase in a model should be treated with caution. Naturally, most interest will focus on low-frequency models, the 'dominant' model of dynamical systems. High-frequency models are also of interest; for instance, in networks such as power electronically controlled electrical transmission systems (O'Reilly, Wood and Osauskas, 2003).

Specifically, Result 3.2 provides a corrected low-frequency model that approximates the system transfer function $G(s, \varepsilon)$ to first-order accuracy for $p = \varepsilon s$ sufficiently small (sufficiently low frequency). Likewise in a dual sense, Result 5.1 provides a corrected high-frequency model that can approximate $G(s, \varepsilon)$ to first-order accuracy for $\varepsilon/p = 1/s$ sufficiently small (sufficiently high frequency). Unlike the corrected low-frequency model, the corrected high-frequency model, however, may not be of reduced order. All $p = \varepsilon s$ and $\varepsilon/p = 1/s$ small approximations are, of course, to the same order of ε approximations in their respective low- and high-frequency scales.

An important property of the first-order corrected low-frequency model in Result 3.2 is that the phase of the original two-frequency-scale system transfer function $G(s, \varepsilon)$ is preserved to within $O(p^2)$ accuracy. This is in contradistinction to the uncorrected low-frequency model, where only the DC-gain part of the high-frequency transfer function is retained (Kokotović, Khalil and O'Reilly, 1999).

Taken together, the results of the chapter provide further strong support for the use of corrected model approximations as opposed to uncorrected model approximations. This applies whether the corrected models are derived in the frequency domain, as here, or they are derived in the time domain using singular perturbation methods (Kokotović, Khalil and O'Reilly, 1999). The implications for the system modelling of transfer-function systems in diverse areas are considerable.

FIVE

MODELS FOR NONLINEAR SINGULARLY PERTURBED SYSTEMS – WHICH TO CHOOSE?

SUMMARY

This chapter addresses the fundamental issue as to which nonlinear system model to choose. The time scale – slow or fast – of primary interest to the modeller is first identified. The second thing to establish is the size of the small positive singular perturbation parameter ε, either known or estimated. Should ε be very small or close to zero, the original singularly perturbed model will be very stiff or ill-conditioned. A simple well-conditioned $\varepsilon = 0$ model beckons. For a very small ε, this simple $\varepsilon = 0$ model, which provides order of ε accuracy to the original singularly perturbed model, is likely to be more than satisfactory. Conversely, should the singular perturbation parameter ε be small but not close to zero, the original singularly perturbed model is not so stiff and, of course, provides perfect accuracy. Why not choose it? The grey area, of course, is where ε is small enough to provoke stiffness problems but not small enough to provide satisfactory order of ε accuracy. In which case, a model corrected in ε with improved accuracy is needed.

5.1 INTRODUCTION

> Alice: "Would you tell me, please, which way I ought to go from here?"
> The Cheshire Cat: "That depends a great deal on where you want to get to."
>
> Alice in Wonderland
> Lewis Carroll

Alice's dilemma is one common to model builders across science and engineering – which model should one choose? At the heart of the dilemma is the purpose that the intended model is expected to serve. Expectations as to what the model might purposely achieve must naturally be tempered by knowledge of the complexities of the phenomenon or system under scrutiny. This modelling dilemma – which model to choose – is at its most acute where the physical phenomenon or engineering system in question possesses dynamics that vary widely in speed or time scale. Such phenomena and systems are commonly described as *singularly perturbed*, and it is to *nonlinear singularly perturbed system models* that this chapter confines its attention.

These nonlinear singularly perturbed system models are usually specified by the system of ordinary differential equations, with initial conditions $x(t_0) = x^0$ and $z(t_0) = z^0$, given by

$$\frac{dx}{dt} = f(x,z,\varepsilon,t), \quad x(t_0) = x^0, \quad x \in R^n \tag{5.1}$$

$$\varepsilon \frac{dz}{dt} = g(x,z,\varepsilon,t), \quad z(t_0) = z^0, \quad z \in R^m \tag{5.2}$$

where $t \geq t_0$ denotes time, and the functions f and g are assumed to be sufficiently many times differentiable in their arguments x, z, ε and t. The parameter ε, the so-called 'singular perturbation parameter', is assumed to be small and positive. It is also assumed, consistent with Chapter 1 and Chapter 2 of Part 1, that the dynamical system (5.1) and (5.2) is *physically realisable* or *causal*.

In theory, there are a number of dynamical system models associated with (5.1) and (5.2) to choose from: at least two models in the slow time scale $t \geq t_0$, and two models in the fast time scale $\tau = (t - t_0)/\varepsilon$, where $\tau \geq 0$. In the slow time scale t of (5.1) and (5.2), the simplest approximate model is, of course, obtained by setting the small parameter ε in (5.2) to zero. The second model in slow time scale t is very familiar, but too often quickly discarded; it is the exact model (5.1) and (5.2) itself. The remaining two models in the fast time scale τ are examined in analogous fashion.

The purpose of this chapter is modest and practical in nature. It is to review the salient features of each model so as to better provide the intending user with guidelines as to any final choice of model or models that they may care to make. In the first instance, this is best done by way of a progressive development of the models, so as to provide the necessary physical insight into the intrinsic two-time-scale nature of the system dynamics as the intending user might view them. Fortunately, this endeavour is greatly assisted by a century of previous mathematical and related work on singular perturbation methods.

Since singular perturbation methods have been so immensely successful in so many fields, this chapter is aimed at a general readership. This will certainly include engineers, model builders in particular, with at least a sideways glance at stiff differential equations in terms of model solution efficiency. The ideas of this chapter, in particular the key notion of separation of time scales, owes much to previous fundamental works. It is consistent with the development of Chapter 3 and Chapter 4.

5.2 THE SINGULAR PERTURBATION INITIAL VALUE PROBLEM

At the start of the twentieth century, fluid mechanics was at an impasse. The classical theory of laminar flow of a fluid past a solid plate held that it was a uniform flow with zero viscosity. From observation and experiment in flow tanks, it was clear enough that uniform flow theory held quite well away from the solid plate. But the flow was decidedly non-uniform in the small region close to the solid plate. In fact, the flow appeared to be practically stationary at the solid plate itself. Something was missing. Whatever it was, it might be small, but it was having a big effect.

The missing element was presented by a twenty-nine-year-old German professor of mechanics named Ludwig Prandtl at the Third International Congress of Mathematicians in Heidelberg in 1904, in a mere seven pages, to immediate and startling effect (Arakeri and Shankar, 2000; Anderson, 2005; Vogel-Prandtl, 2004). That missing element from the classical theory was the thin boundary layer next to the plate where at small but non-zero viscosity the fluid flow rapidly increased from zero velocity next to the plate to a value close to that of the uniform flow further out from the plate.

The boundary layer was called the inner solution where there was a rapid change in flow velocity, the so-called 'boundary layer jump'. Next to the boundary layer, further into the fluid flow, was the outer solution where the change in flow velocity was much more gradual. Adding the inner solution

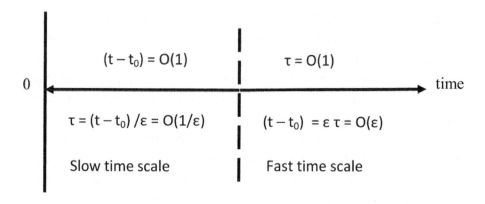

Figure 5.1 *The two time scales (t – t_0) and τ*

(the new bit) to the outer solution (the classical bit) yielded the final uniform solution approximation that the classical theory had so signally failed to provide. This little seven-page 1904 paper became the foundation of modern fluid mechanics, and its sister discipline of aerodynamics with its lift and drag (Anderson, 2005). It also signalled the birth of singular perturbation theory (O'Malley, 2010).

The inner solution and outer solution that Prandtl introduced satisfy the uniform state vector approximations to within $O(\varepsilon)$ given by

$$x(t,\varepsilon) = X_0(t) + O(\varepsilon)$$ (5.3)

$$z(t,\varepsilon) = Z_0(t) + \hat{z}_0(\tau) + O(\varepsilon)$$ (5.4)

where $X_0(t) \equiv X(t, 0)$ and $Z_0 \equiv Z(t, 0)$ denote the outer solutions in slow time $t \geq 0$, and $\hat{z}_0(\tau) \equiv \hat{z}(\tau, 0)$ denotes the inner or boundary layer correction solution in stretched or fast time $\tau = (t - t_0)/\varepsilon$, where $\tau \geq 0$ ($\tau = 0$ for $t = t_0$). Both the slow time scale $t \geq t_0$ and the fast time scale $\tau \geq 0$ are as depicted in Figure 5.1. In the slow time scale $t \geq t_0$, the boundary layer solution $\hat{z}_0(\tau)$ in (5.4) is of short duration $\tau = \varepsilon(t_1 - t_0)$ for some $t_1 > t_0$, which is $O(\varepsilon)$.

The corresponding differential equation model in slow time scale $t \geq t_0$, as in (5.1) and (5.2), assumes the singularly perturbed form.

$$\frac{dx}{dt} = f(x, z, \varepsilon, t), \quad x(t_0) = x^0, \quad x \in R^n$$ (5.5)

$$\varepsilon\frac{dz}{dt} = g\left(x,z,\varepsilon,t\right), \qquad z\left(t_0\right) = z^0, \qquad z \in R^m \tag{5.6}$$

Equations (5.5) and (5.6) are asymmetrical in nature due to the asymmetrical position of the small parameter ε in the system of differential equations. Inspection of (5.6) reveals that $z(t)$ bears the full brunt of the $1/\varepsilon$ large derivative dz/dt, giving rise to a boundary layer jump $\hat{z}_0\left(\tau\right)$ in $z(t)$, as in (5.4). Inspection of (5.5) reveals that $x(t)$ is less affected, as in (5.3), at least to within $O(\varepsilon)$. In other words, equation (5.5) acts like a slow or low-pass filter on the rather more dynamic $z(t)$ from equation (5.6).

This can be seen to better effect if the singularly perturbed model (5.5) and (5.6) is rewritten in the fast time scale $\tau = (t - t_0)/\varepsilon$, where $\tau \geq 0$ as

$$\frac{dx}{d\tau} = \varepsilon f\left(x,z,\varepsilon,\tau\right), \qquad x\left(0\right) = x^0, \qquad x \in R^n \tag{5.7}$$

$$\frac{dz}{d\tau} = g\left(x,z,\varepsilon,\tau\right), \qquad z\left(0\right) = z^0, \qquad z \in R^m \tag{5.8}$$

With that small parameter ε multiplying the function f, equation (5.7) depicts a slowly varying system that has an averaging or low-pass filtering effect on the faster system (5.8).

Let us now consider the *outer* solutions $X_0(t) \equiv X(t, 0)$ and $Z_0 \equiv Z(t, 0)$ of (5.3) and (5.4); in particular, where they come from. As expected, the outer solutions come from setting $\varepsilon = 0$ in the original singularly perturbed system of differential equations (5.5) and (5.6) in the slow time scale $t \geq t_0$, omitting or ignoring the initial condition on z, to obtain

$$\frac{dX_0\left(t\right)}{dt} = f\left(X_0\left(t\right),Z_0\left(t\right),0,t\right), \qquad X_0\left(t_0\right) = x^0 \tag{5.9}$$

$$0 = g\left(X_0\left(t\right),Z_0\left(t\right),0,t\right) \tag{5.10}$$

Equation (5.10) is an algebraic or transcendental equation. It is here that we make our first major assumption that there exists at least one root $Z_0(t) = Z(t, 0)$ of the algebraic equation (5.10).

Assumption 2.1: In a domain of interest, there exists at least one real root of equation (5.10) given by

$$Z_0(t) = \varphi\left(X_0(t), t \right) \tag{5.11}$$

Assumption 2.1 ensures that, to each real root of (5.10), there corresponds a well-defined reduced-order dynamical model in $X_0(t) \equiv X(t, 0)$. This reduced-order model in $X_0(t)$ is obtained by substituting (5.11) into (5.9) to arrive at

$$\frac{dX_0(t)}{dt} = f\left(X_0(t), \varphi\left(X_0(t), t \right), 0, t \right), \quad X_0(t_0) = x^0 \tag{5.12}$$

There may be more than one real root (5.11) of (5.10) of course. To each such root (5.11), there is a corresponding reduced-order model (5.12).

Let us now consider the *inner* solution $\hat{z}_0(\tau) \equiv \hat{z}(\tau, 0)$ of (5.3) and (5.4); in particular, where it comes from. As expected, the inner solution $\hat{z}_0(\tau)$ arises as a boundary layer correction $\hat{z}_0(\tau) = z - Z_0$ to the outer solution Z_0 of the previous step. The inner solution $\hat{z}_0(\tau)$ is thereby got from setting $\varepsilon = 0$ in the original singularly perturbed system of differential equations (5.7) and (5.8) exhibited in the fast time scale $\tau \geq 0$ to obtain

$$\frac{dx(\tau)}{d\tau} = 0, \quad x(0) = x^0 \tag{5.13}$$

$$\frac{d\hat{z}_0(\tau)}{d\tau} = g\left(x^0, \hat{z}_0(\tau) + Z_0(t_0), 0, t_0 \right), \quad \hat{z}_0(0) = z^0 - Z_0(t_0) \tag{5.14}$$

Note that t_0 is a fixed parameter in (5.14). Also, from (5.13), it is observed that $x(\tau) = \text{constant} = x_0$, a fixed parameter in the fast time scale $\tau \geq 0$. It is here that we make our second assumption, that the well-defined reduced-order boundary layer system (5.14) is asymptotically stable such that $\hat{z}_0(\tau)$ decays to zero as τ tends to infinity within the boundary layer.

Assumption 2.2: The eigenvalues of the Jacobian matrix $\partial g/\partial z$ in (5.14) have negative real parts.

Assumption 2.2 also implies that the Jacobian matrix $\partial g/\partial z$ is non-singular, thereby ensuring that there exists at least a local unique solution $Z_0(t) = \varphi(X_0(t), t)$ of (5.11) to the earlier outer solution equation (5.10).

Finally, in the initial value solution (5.4) it is necessary that the final value of the inner term $\hat{z}_0(\tau) = \hat{z}(\tau,0)$ as τ tends to infinity agrees with the initial value of the outer term $Z_0(t) = Z(t, 0)$ as t tends to t_0. Therefore, the matching condition of inner term to outer term is $\hat{z}_0(\infty) = Z_0(t_0)$.

Under the assumptions of there being sufficient differentiability of function arguments, Assumption 2.1 and Assumption 2.2, the initial value uniform $O(\varepsilon)$ state vector approximations (5.3) and (5.4) hold for the singularly perturbed system (5.5) and (5.6). In fact, under these same assumptions, the uniform $O(\varepsilon)$ state approximations (5.3) and (5.4) generalise to the following initial value uniform $O(\varepsilon^{N+1})$ state approximations.

$$x(t,\varepsilon) = X(t,\varepsilon) + \varepsilon\hat{x}(\tau,\varepsilon) + O(\varepsilon^{N+1}) \tag{5.15}$$

$$z(t,\varepsilon) = Z(t,\varepsilon) + \hat{z}(\tau,\varepsilon) + O(\varepsilon^{N+1}) \tag{5.16}$$

where the outer solutions $X(t, \varepsilon)$ and $Z(t, \varepsilon)$, and the inner solutions $\hat{x}(\tau,\varepsilon)$ and $\hat{z}(\tau,\varepsilon)$, are respectively defined by the power series expansions

$$X(t,\varepsilon) = \sum_{j=0}^{N} X_j(t)\varepsilon^j, \quad Z(t,\varepsilon) = \sum_{j=0}^{N} Z_j(t)\varepsilon^j \tag{5.17}$$

$$\hat{x}(\tau,\varepsilon) = \sum_{j=0}^{N-1} \hat{x}_j(\tau)\varepsilon^j, \quad \hat{z}(\tau,\varepsilon) = \sum_{j=0}^{N} \hat{z}_j(\tau)\varepsilon^j \tag{5.18}$$

Matched asymptotic expansions of the form of (5.17) and (5.18) were presented by Prandtl himself in the winter of 1931–1932 (O'Malley, 2010). The result (5.15) and (5.16) is known as the Tikhonov-Levinson theorem or simply the Tikhonov theorem (Tikhonov, 1948, 1952; Levinson, 1950; Leven and Levinson, 1954), and is proved using Taylor series expansions under the weakest conditions by Hoppensteadt (1971). There are many versions of this result from slightly different perspectives; for instance, see Vasil'eva (1963), O'Malley (1971, 1974) and Slavona (1995). Engineers in fluid mechanics have used Kaplun (1967) and the more popular Van Dyke (1964) to successively develop uniform solution approximations based upon matched asymptotic expansions of the form of (5.17) and (5.18). Others from a mathematical background have used Wasow (1965). Even with the help of symbolic computing, these matched asymptotic expansions can be tricky. Models based upon such expansions are beyond the scope of this chapter. The development presented here, of what is now the classical theory of singular perturbations, is however important from a conceptual point of view and is useful background to the consideration of models in Section 5.3 and Section 5.4.

5.3   COMPUTATIONAL   EFFICIENCY,   NUMERICAL STABILITY AND THE SINGULAR PERTURBATION PARAMETER: STIFFNESS

By computational efficiency and numerical stability in the solution of ordinary differential equations, we are referring to the stiffness or otherwise of the system of equations describing the physical dynamical system model (5.5) and (5.6). While the stiffness of ordinary differential equations has been studied extensively since 1952 (Curtiss and Hirschfelder, 1952), even fifty years later (Spijker, 1996), stiffness is referred to as a general phenomenon rather than as a precise solution property.

Nonetheless, with regard to the numerical solution of $dz/dt = g/\varepsilon$ in (5.6) in the inner or boundary layer region, where the solution $z(t)$ undergoes rapid change, it is to be expected that the integration step size h needs to be relatively small if it is to adequately describe these fast dynamics. On the other hand, with regard to the numerical solution of $dx/dt = f$ in (5.5) in the outer region, where the solutions $x(t)$ and $z(t)$ undergo slow change, it is to be expected that the integration step size h can be relatively large. Such a system of differential equations is described as 'stiff' where the integration step size h is forced to be unacceptably small in a region where the solution is varying relatively slowly.

While there are many definitions of stiffness in the literature, the following definition of Spijker (1996) is representative.

Definition 3.1: Suppose L is the Lipschitz constant of F as defined by the smallest value of L such that

$$\left|F\left(t,y\right)-F\left(t,x\right)\right|\leq L\left|y-x\right| \qquad (5.19)$$

for all $t \in [t_0, T]$ and $x, y \in D \subset R^s$.

Then, the differential equation $dx\,/\,dt = F\left(t,x\right)$ is said to be stiff if the Lipschitz constant of F and the integration step size h satisfy

$$hL \gg 1 \qquad (5.20)$$

Loosely speaking, the Lipschitz condition (5.19) of Definition 3.1 informs us that the rate of change of $F(x)$ with respect to x is not more than L. Consequently, the differential equation $dx/dt = F(x)$ is considered stiff if the product of L and the small integration step size h remains stubbornly large, as defined by (5.20).

In the context of the singularly perturbed system of differential equations (5.5) and (5.6), dividing the second equation (5.6) through by ε, one has that the $1/\varepsilon$ large term enters the associated Lipschitz constant of g/ε. Therefore, by Definition 3.1, we arrive at the following result.

Result 3.1: The smaller the singular perturbation parameter ε is, the stiffer the singularly perturbed system of differential equations (5.5) and (5.6) is for a given integration step size h in the sense of Definition 3.1.

Result 3.1 is little more than a formal restatement of the widely known fact (O'Malley, 1988) that the more singularly perturbed or ill-conditioned the system is, the stiffer the system is with regard to numerical integration. It does, however, serve notice that the size of the small singular perturbation parameter ε, either known or estimated, is of critical importance to our choice of appropriate models, at least as far as stiffness is concerned. The importance of the size of the singular perturbation parameter ε in terms of achieving a suitable level of model approximation has already been encountered in Chapter 4.

The relative stiffness of a singularly perturbed system is illustrated by two very different examples. These two examples will serve as exemplars, both here in our discussion of stiffness and in Section 5.4 on the best choice of models. The first example from the chemical sciences is the opening example of O'Malley (1988) on the very subject of stiffness.

Example 3.1: Enzyme kinetics (Murray, 1977; O'Malley, 1988)

In the Michaelis-Menton theory of enzyme kinetics (Murray, 1977), the non-dimensional substrate concentration $x(t)$ and the non-dimensional complex concentration $z(t)$ satisfy the following nonlinear singularly perturbed system of differential equations

$$\frac{dx}{dt} = -x + \left(x + \kappa - \lambda\right)z, \qquad x(0) = 1 \tag{5.21a}$$

$$\varepsilon\frac{dz}{dt} = x - \left(x + \kappa\right)z, \qquad z(0) = 0 \tag{5.21b}$$

where κ and λ are positive constants. The singular perturbation parameter ε, the initial ratio of enzyme to substrate concentrations, is typically of order 10^{-6} and is extremely small. Consequently, it is fair to say that the singularly perturbed system (5.21) is stiff in the sense of Definition 3.1, with attendant numerical solution difficulties.

Example 3.2: The synchronous machine (Anderson and Fouad, 1977; Kokotović, Khalil and O'Reilly, 1999)

A benchmark representation of power systems is the synchronous machine connected to an infinite bus through a transmission line. The synchronous generator is represented by the one-axis model (Anderson and Fouad, 1977; Kokotović, Khalil and O'Reilly, 1999) given by

$$\varepsilon = 1 / \tau_{d0} = 0.1515$$

$$\tau_{d0}\frac{dE_q}{d\tau} = \frac{x_d + x_e}{\acute{x}_d + x_e}E_q + \frac{x_d - \acute{x}_d}{\acute{x}_d + x_e}V\cos\delta + E_{FD} \tag{5.22a}$$

$$\frac{d\delta}{d\tau} = \omega \qquad (5.22b)$$

$$M\frac{d\omega}{d\tau} = -D\omega + P_m - \frac{1}{x_d' + x_e} E_q V \sin\delta \qquad (5.22c)$$

where

τ_{d0} = direct-axis open-circuit transient time constant

x_d = direct-axis synchronous reactance of the generator

x_d' = direct-axis transient reactance of the generator

E_q = instantaneous voltage proportional to field-flux linkage

δ = angle between voltage of infinite bus and $\dot{E}_q$

M = inertial constant of the generator

D = damping coefficient of the generator

E_{FD} = field voltage (assumed constant)

P_m = mechanical input power to the generator (assumed constant)

ω = slip velocity of the generator

x_e = reactance of the transmission line

V = voltage of the infinite bus

The variable E_q is usually slower than δ and ω as a result of a relatively large value of the time constant τ_{d0} in (5.22a). Consequently, E_q is viewed as a slow variable and δ and ω are viewed as fast variables in a singularly perturbed system model described by

$$\frac{dx}{dt} = -ax + b\left[\cos\left(z_1 + \overline{\delta}\right) - \cos\overline{\delta}\right] \triangleq f(x,z) \qquad (5.23a)$$

$$\varepsilon\frac{dz_1}{dt} = z_2 \triangleq g_1(x,z) \qquad (5.23b)$$

$$\varepsilon\frac{dz_2}{dt} = -\lambda z_2 - c\left[(1+x)\sin\left(z_1 + \overline{\delta}\right) - \sin\overline{\delta}\right] \triangleq g_2(x,z) \qquad (5.23c)$$

where the state variables are taken as $x = \left(E_q - \overline{E}_q\right)/\overline{E}_q$, $z_1 = \delta - \overline{\delta}$ and $z_2 = \omega - \overline{\omega}$ with the barred quantities denoting their equilibrium values. The singular perturbation parameter $\varepsilon = 1/\tau_{d0} = 0.1515$ in the new time scale $t = \tau / \tau_{d0}$.

Note that the value of singular perturbation parameter $\varepsilon = 0.1515$ of the synchronous machine model described by (5.23) is of the order of 10^5 larger than the value of ε in the enzyme kinetics model of Example 3.1. Values of ε of the order of $\varepsilon = 0.1$, such as in the synchronous machine model considered here, are representative of all electrical machines. The singular perturbation parameter ε is small but it is not very small or close to zero. Unlike Example 3.1, it is fair to say that the singularly perturbed system (5.23) is not stiff in the sense of Definition 3.1, and is without attendant numerical solution difficulties.

This discussion on the relative stiffness of different singularly perturbed models, as defined by the smallness of the singular perturbation parameter ε, is extremely helpful in Section 5.4 on making the best choice of model out of those models available in a particular time scale. Naturally, one has a preference for models that are well conditioned; that is to say, they present no stiffness issues. These stiffness issues are more prominent in the physical and chemical sciences (for example, Lebon and Rizzoni, 2010; Fernando, Oliveira and Fernando, 2009; Kumar, 2011) where the singular perturbation parameter ε is often of order 10^{-2} or less, as in Example 3.1. Stiffness issues are less of a problem in engineering (for example, Williams, 2008; Lakrad and Belhaq, 2002; Kokotović, Khalil and O'Reilly, 1999), where the singular perturbation parameter ε, while still small, may be 0.1 or greater, as in Example 3.2. Indeed, in Chapter 3 and Chapter 4, the scaling parameter ε could be at the modeller's discretion, where in theory, ε might be increased to values unheard of in the sciences. In some cases, one might reasonably question the validity of attendant engineering model approximations.

5.4   THE BEST CHOICE OF SYSTEM MODEL

This narrative has come some distance in preparing an answer to Alice's basic question of 'Which way to go?' in relation to the best choice of system model the engineer or others might make. As in Chapter 3 and Chapter 4, it is critically important to establish from the outset which particular time scale is of major interest, while of course retaining a subsidiary interest in the other time scale as it affects the principal time-scale model chosen. Starting with the *slow* time scale

t, where $(t - t_0) = \varepsilon \tau$ is $O(1)$, let us first re-examine the two slow models available to us from Section 5.2. This will be followed by a re-examination of the two *fast* models from Section 5.2 in the fast time scale $\tau = (t - t_0)/\varepsilon$, where τ is $O(1)$.

Once the requisite time scale of interest has been specified, the next important question, as we shall see, is to consider the known or estimated value of the singular perturbation parameter ε and to ask is ε very small or merely small? The issue of the smallness of the singular perturbation parameter ε is of great importance in making the most appropriate selection of system model. And so, let us move on to the slow-time-scale model candidates to be considered.

5.4.1 THE SLOW-TIME-SCALE MODEL CANDIDATES

In the slow time scale t, where $(t - t_0) = \varepsilon \tau = O(1)$, the first model is as follows.

The uncorrected slow $\varepsilon = 0$ model:

$$\frac{dx}{dt} = f\left(x, \varphi(x,0), t\right), \quad x\left(t_0\right) = x^0, \quad x \in R^n \tag{5.24}$$

As we saw in Section 5.2, this is the simplest slow model available to us, and it is obtained by setting $\varepsilon = 0$ in the original singularly perturbed system model (5.5) and (5.6). This model is well conditioned, with none of the stiffness issues raised in Section 5.3. It is also of lower order n than the original model (5.5) and (5.6), which is of order $n + m$; this is useful should the original system be of high order.

The uncorrected model (5.24) provides an $O(\varepsilon)$ approximation for $x(t)$, for $t \geq t_0$, to the original singularly perturbed system (5.5) and (5.6) in the designated slow time scale $(t - t_0) = \varepsilon \tau = O(1)$. The initial fast transient from initial time t_0 to time t_1 is but of the short $(t_1 - t_0) = \varepsilon \tau = O(\varepsilon)$ duration for the original singularly perturbed model (5.5) and (5.6) in this slow time scale. In the uncorrected slow model (5.24), this short transient is ignored by setting $\varepsilon = 0$. This is commonly done to good effect by modellers with no knowledge of singular perturbation techniques.

Should an $O(\varepsilon)$ approximation for $z(t)$, for $t \geq t_0$, also be desired, a boundary layer correction $\hat{z}_0\left(\tau\right)$, as in equation (5.4), is required.

How good the model approximation (5.24) actually is will crucially depend upon the smallness of the singular perturbation parameter ε. It needs to be very small. Should the parameter ε not be so small, the approximations (5.3) and (5.4) may be insufficiently accurate, and a higher-order matched expansion along the lines of (5.15) and (5.16) may be considered. This could be significant when the singular perturbation parameter ε is not that small, as in Example 3.2.

An alternative choice of model is the following.

The original singularly perturbed model in slow time scale t:

$$\frac{dx}{dt} = f\left(x,z,\varepsilon,t\right), \quad x\left(t_0\right) = x^0, \quad x \in R^n \tag{5.25}$$

$$\varepsilon\frac{dz}{dt} = g\left(x,z,\varepsilon,t\right), \quad z\left(t_0\right) = z^0, \quad z \in R^m \tag{5.26}$$

The second choice of model in the designated slow time scale $(t - t_0) = \varepsilon\tau$ of $O(1)$ is the original singularly perturbed model itself. This model may be ignored by some in the rush to obtain new theoretical results, but not by model builders. After all, that original model (5.25) and (5.26) may be the fruit of much hard modelling and scaling labour. It has decided advantages. The first advantage is that the model with initial conditions is available. The second advantage is that the model is exact; no approximations are called for.

So how would this model (5.25) and (5.26) run? For a simulation or solution, a variable-size integration step is required, since a small-size integration step is needed to capture the fast transient over the initial short interval $(t_1 - t_0) = \varepsilon\tau = O(\varepsilon)$, while a larger-size integration step suffices to capture the slow dynamics once the fast dynamics have decayed to zero at time $t > t_1$. An example of an application of this solution procedure is the simulation of a power system with electromagnetic and electromechanical dynamics in separate time scales (Pöller and Schmieg, 1997).

As predicted in Section 5.3, the singularly perturbed system model (5.25) and (5.26) comes with a major disadvantage. In contradistinction to the approximate model (5.24), the model (5.25) and (5.26) is stiff for very small values of the singular perturbation ε. This ceases to be such a problem when ε is not so very small. Also, the model is of full order for all ε. Let us recap on the two candidate slow models by way of an example.

Example 4.1: Example 3.1 and Example 3.2, respectively, exhibit a very small value of ε at $O(10^{-6})$ and a not-so-small value of ε at 0.1515. These very different values of ε will largely determine the most promising slow-model candidates.

In the former Example 3.1 case, the very small ε eliminates the full-order model from consideration due to stiffness issues. In the latter Example 3.2 case, ε is not so small, the full-order model is not so stiff, and it should run adequately.

The ideal model for Example 3.1 is the simple uncorrected one since $O(10^{-6})$ accuracy surely suffices. If not, a corrected model will provide $O(10^{-12})$ accuracy.

A possible model for Example 3.2 is a corrected model that provides $O(0.1515^2) = O(0.0230)$ accuracy. The uncorrected model (5.24) is likely to be insufficiently accurate at $O(0.1515)$ accuracy. The full-order model (5.25) and (5.26) is however exact and surely remains a contender.

The fast-time-scale model candidates are now considered.

5.4.2 THE FAST-TIME-SCALE MODEL CANDIDATES

In the fast time scale $\tau = O(1)$, where $\tau = (t - t_0)/\varepsilon$, the first of two models is as follows.

The uncorrected fast $\varepsilon = 0$ model:

$$\frac{dz}{d\tau} = g\left(x^0, z, \tau\right), \qquad z\left(0\right) = z^0 \tag{5.27}$$

This is the simplest fast model available to us and is obtained by setting $\varepsilon = 0$ in the original singularly perturbed system model (5.5) and (5.6) in the fast time scale $\tau = O(1)$. Like the uncorrected slow model (5.24), this uncorrected fast model (5.27) is well conditioned, with none of the stiffness issues raised in Section 5.3. It is also of lower order m than the original model (5.5) and (5.6) which is of order $n + m$; this is useful should the original system be of high order.

The uncorrected model (5.27) provides an $O(\varepsilon)$ approximation to the original singularly perturbed system (5.5) and (5.6) in the designated fast time scale $\tau = (t - t_0)/\varepsilon = O(1)$. The slow state x is frozen at its initial value $x = x^0$ in the original singularly perturbed model (5.5) and (5.6) in this fast time scale τ at $\varepsilon = 0$.

Again, how accurate the model approximation (5.27) actually is will crucially depend upon the smallness of the singular perturbation parameter ε. It needs to be very small. Should the parameter ε not be so small, the approximation (5.27) is not so accurate, and in the absence of a suitable corrected nonlinear fast model, the original system model needs to be considered.

The original singularly perturbed model in fast time scale τ:

$$\frac{dx}{d\tau} = \varepsilon f\left(x,z,\varepsilon,\tau\right), \quad x\left(0\right)=x^0, \quad x \in R^n \qquad (5.28)$$

$$\frac{dz}{d\tau} = g\left(x,z,\varepsilon,\tau\right), \quad z\left(0\right)=z^0, \quad z \in R^m \qquad (5.29)$$

The original singularly perturbed model (5.28) and (5.29) enjoys the advantages of being available and exact; no approximations are called for. The model runs in the fast time scale τ with small integration step sizes.

A disadvantage of the original model (5.28) and (5.29) is that it is stiff for very small values of the singular perturbation ε. This ceases to be such a problem when ε is not so very small. Also, the model is of full order for all ε. Let us conclude by continuing with our example.

Example 4.2: As noted previously, Example 3.1 and Example 3.2 respectively exhibit a very small value of ε at $O(10^{-6})$ and a not so small value of ε at 0.1515.

In the former Example 3.1 case, the very small value of ε eliminates the original full-order model from consideration due to stiffness issues but the uncorrected fast model at $O(10^{-6})$ accuracy should be sufficient.

In the latter Example 3.2 case, ε is not so small, the original full-order model is not so stiff, and it should run adequately. The uncorrected model is likely to be insufficiently accurate at $O(0.1515)$ accuracy.

One may wonder why a corrected fast model is not considered as well, as a halfway house in accuracy and complexity between the uncorrected fast model (5.27) and the original model (5.7) and (5.8) in the fast time scale. A candidate fast corrected model does present itself when one realises that a slow state x is required that is rather more than the frozen state x^0 in the uncorrected case so as to be of a slowly varying form not unlike x in (5.7) of the full model. One possibility for a fast corrected model is described by

$$\frac{dx}{d\tau} = \varepsilon f\left(x, z^0, \tau\right), \qquad x(0) = x^0, \quad x \in R^n \tag{5.30}$$

$$\frac{dz}{d\tau} = g\left(x, z, \tau\right), \qquad z(0) = z^0, \quad z \in R^m \tag{5.31}$$

Whatever the merits of the candidate corrected fast model (5.30) and (5.31) *vis-à-vis* accuracy of approximation, the model is still singularly perturbed, like the full model (5.7) and (5.8), and therefore stiff for very small ε. It offers no advantages over the original exact model (5.7) and (5.8) in the fast time scale τ. So, it is omitted from further consideration.

In summary, one observes an asymmetry between the set of two slow models and the set of two fast models. As mentioned previously, this asymmetry arises from the asymmetry of the original singularly perturbed system itself in that ε multiplies some but not all of the state variables in (5.5) and (5.6).

5.5   CONCLUSIONS

The question of which model to choose is as fundamental for nonlinear singularly perturbed systems as it is for any other dynamical system. And yet there appears to be little guidance on the issue of model choice in the literature except for the occasional paper in a specific application area. How can this be? The answer is twofold.

Firstly, the choice of model is ultimately tied to the purpose of the model the model builder has in mind. The needs of the aerodynamicist studying air flow past an aerofoil will be very different from the control specialist intent upon meeting control objectives. One model will need to be detailed, the other less so. Like when someone is intending to marry, there would appear to be little incentive in trying to offer general guidance on what is, after all, a personal matter.

Then there is the second reason as to why this fundamental issue has been conveniently sidestepped in the literature. The conclusions as to model choice may be far reaching in nature, but their very simplicity and limited warrant are hardly the stuff of contemporary research papers.

Nonetheless, it is heartening to read O'Malley (1988), which in an uncomplicated way shows how the study of singular perturbations is intertwined with that of stiff differential equation models. In a sense, this chapter continues that discussion, but directed at the question as to what model to choose.

The conclusions of this third and final chapter of Part 2 are therefore simple and practical. Firstly, one must identify the time scale, slow or fast, of most interest in the model. Secondly, one must determine the size of the small positive singular perturbation parameter ε; is it very small (close to zero) or merely small but not close to zero? In the former case, a simple well-conditioned uncorrected $\varepsilon = 0$ model may well suffice. In the latter case, the original singular perturbation model may not present undue stiffness problems and the modeller proceeds with it as with a bird in the hand. Hard-pressed modellers need no urging: go for the low-hanging fruit.

The aforementioned simple general conclusions serve a large class of singularly perturbed nonlinear system models: those with a very small ε and those with a not-so-small ε. Inevitably, however, there is a grey area in which ε is sufficiently small to present stiffness problems in the original singular perturbation model, but not small enough that the uncorrected $\varepsilon = 0$ model with its accuracy of $O(\varepsilon)$ is accurate enough. A corrected model that provides $O(\varepsilon^2)$ or even higher accuracy may be required. There is nothing for it, then, but to tackle those matched asymptotic expansions referred to at the end of Section 5.2.

One final thing. A shortcoming of this book is the lack of numerical examples. A cursory inspection of block-diagram simulation packages, such as Matlab and Simulink, reveals them to be straightforward and fun to use. The reader should therefore enjoy supplementing the text with exploratory numerical examples of their own.

All of this brings this book to a sort of conclusion. In that, I am reminded of the great jazz saxophonist John Coltrane allegedly confiding in his colleague Miles Davis that he was having great difficulty in bringing his increasingly long solos to a conclusion. The reply was, "How about taking the horn out of your mouth?"

EPILOGUE TO PART 2

The subject matter of Part 2 on singularly perturbed systems is quite different from that of Part 1 on dynamical systems. Nonetheless, Part 2 has benefitted immensely from an emerging view in Part 1 as to what, at its core, engineering with its physical narrative is. Part 2 on singular perturbations and two-frequency-scale systems provides further compelling evidence that such a physical narrative can – indeed, must – yield important new results and new research directions.

REFERENCES FOR PART 2

Anderson, J. D. (2005). Ludwig Prandtl's boundary layer. *Physics Today, 58,* 42–48.

Anderson, P. M. and Fouad, A. A. (1977). *Power System Control and Stability.* Iowa State University Press.

Appleton, E. V. (1923). The automatic synchronization of triode oscillators. *Proceedings of the Cambridge Philosophical Society, 21,* 231–248.

Arakeri, J. H. and Shankar P. N. (2000). Ludwig Prandtl and boundary layers in fluid flow. *Resonance, 5,* 48–53.

Bagirov, N., Vasil'eva, A. B., and Imanaliev, M. I. (1967). The problem of asymptotic solutions of automatic control problems. *Differential Equations, 3,* 985–988.

Biernson, G. (1988). *Principles of Feedback Control – Volume 1.* Wiley Interscience, New Jersey.

Bode, H. W. (1940). Relations between attenuation and phase in feedback amplifier design. *Bell System Technical Journal, 19,* 421–454.

Bode, H. W. (1945). *Network Analysis and Feedback Amplifier Design.* Van Nostrand, Princeton, New Jersey.

Cavalcanti, A. L. O., Kienitz, K. H., and Kadirkamanathan, V. (2018). Identification of two-time scaled systems using prefilters. *Journal of Control Science and Engineering,* Article ID 3138149.

Chaplais, F. and Alaoui, K. (1996). Two-time scaled parameter identification by coordination of local identifiers. *Automatica, 32,* 1303–1309.

Curtiss, C. F. and Hirschfelder, J. O. (1952). Integration of stiff equations. *Proceedings of the National Academy of Sciences, 38,* 235–243.

Fernando, M. P., Oliveira, A. A. M., and Fernando, F. F. (2009). Asymptotic analysis of stationary adiabatic premixed flames in porous inert media.

Combustion and Flame, 156, 152–165.

Francke, M., Pogromsky, A., and Nijmeijer, H. (2020). Huygen's clocks: 'sympathy' and resonance. *International Journal of Control, 93,* 274–281.

Friis, H. T. and Jensen, A. G. (1924). High frequency amplifiers. *Bell System Technical Journal, 3,* 181–205.

Hoppensteadt, F. (1971). Properties of solutions of ordinary differential equations with small parameters. *Communications on Pure and Applied Mathematics, 34,* 807–840.

Hu, Y. and Wang, Y-Y. (2015). Two time-scaled model identification with application to battery state estimation. *IEEE Transactions on Control Systems Technology, 23,* 1180–1188.

Kaplun, S. (1967). *Fluid Mechanics and Singular Perturbations.* Academic Press, New York.

Kevorkian, J. and Cole, J. D. (1996). *Multiple Scale and Singular Perturbation Methods.* Springer, Berlin.

Kokotović, P. V. and Khalil, H. K. (Eds.) (1986). *Singular Perturbations in Systems and Control* (Reprint). IEEE Press, New York.

Kokotović, P. V., Khalil, H. K., and O'Reilly, J. (1999). *Singular Perturbation Methods in Control: Analysis and Design.* SIAM Classics in Applied Mathematics, Philadelphia.

Kokotović, P. V. and Sannuti, P. (1968). Singular perturbation method for reducing the model order in optimal control design. *IEEE. Transactions on Automatic Control, 13,* 377–384.

Kumar, M. P. (2011). Methods for solving singular perturbation problems arising in science and engineering. *Mathematics and Computer Modelling, 54,* 556–575.

Lakrad, F. and Belhaq, M. (2002). Periodic solutions of strongly non-linear oscillators by the multiple scales method. *Journal of Sound and Vibration, 258,* 677–700.

Lebon, F. and Rizzoni, R. (2010). Asymptotic analysis of a thin interface, the case involving similar rigidity. *International Journal of Engineering Science, 48,* 473–486.

Leven, J. J. and Levinson, N. (1954). Singular perturbations of non-linear systems of differential equations and an associated boundary layer equation. *Journal of Rational Mechanics and Analysis, 3,* 247–270.

Levinson, N. (1950). Perturbations of discontinuous solutions of non-linear systems of differential equations. *Acta Mathematica, 82,* 71–106.

Luse, D. (1986). Frequency domain results for systems with multiple time scales. *IEEE Transactions on Automatic Control*, *31*, 918–924.

Luse, D. (1988). State-space realization of multiple-frequency-scale systems. *IEEE Transactions on Automatic Control*, *33*, 185-187.

Luse, D. and Khalil, H. K. (1985). Frequency domain results for systems with slow and fast dynamics. *IEEE Transactions on Automatic Control*, *30*, 1171-1179.

MacFarlane, A. G. J. (1979). The development of frequency-response methods in automatic control. *IEEE Transactions on Automatic Control*, *24*, 250–265.

Murray, J. D. (1977). *Lectures on Nonlinear Differential-Equation Models in Biology*. Clarendon Press, Oxford.

Oloomi, M. H. and Sawan, M. E. (1995). Characterization of zeros in two-frequency-scale systems. *IEEE Transactions on Automatic Control*, *40*, 92–95.

Oloomi, M. H. and Shafai, B. (2004). Two-time-scale distributions and singular perturbations. *International Journal of Control*, *77*, 1040–1049.

O'Malley, R. E. (1971). Boundary layer methods for nonlinear initial value problems. *SIAM Review*, *13*, 425–434.

O'Malley, R. E. (1974). *Introductions to Singular Perturbations*. Academic Press, New York.

O'Malley, R. E. (1988). On nonlinear singularly perturbed initial value problems. *SIAM Review*, *30*, 193–212.

O'Malley, R. E. (2010). Singular perturbation theory: a viscous flow out of Göttingen. *Annual Review of Fluid Mechanics*, *42*, 1–17.

O'Malley, R. E. (2014). *Historical Developments in Singular Perturbations*. Springer, Berlin.

O'Reilly, J. (1986). Robustness of linear feedback control systems to unmodelled high-frequency dynamics. *International Journal of Control*, *44*, 1077–1088.

O'Reilly, J., Wood, A. R., and Osauskas, C. (2003). Frequency domain based control design for an HVdc converter connected to a weak AC network, *IEEE Transactions on Power Delivery*, *18*, 1028–1033.

Peponides, G., Kokotović, P. V., and Chow, J. H. (1982). Singular perturbations and time-scales in non-linear models of power systems. *IEEE Transactions on Circuits and Systems*, *29*, 758–767.

Pöller, M. and Schmieg, M. (1997). The efficient simulation of multiple time scale systems. *International Conference on Power System Transients*, Seattle, 17–22.

Porter, B. and Shenton, A. T. (1975). Singular perturbation analysis of the transfer function matrices of a class of multivariable linear systems. *International Journal of Control, 21*, 655–660.

Shaker, H. R. (2009). Frequency-domain generalized singular perturbation method for relative error model order reduction. *Journal of Control Theory and Applications, 7*, 57–62.

Slavona, A. (1995). Nonlinear singularly perturbed systems of differential equations: A survey. *Mathematical Problems in Engineering, 1*, 275–301.

Söderström, T. and Stoica, P. (1989). *System Identification*. Prentice Hall, New York.

Spijker, M. N. (1996). Stiffness in numerical initial-value problems. *Journal of Computational and Applied Mathematics, 72*, 393–406.

Tikhonov, A. N. (1948). On the dependence of the solution of differential equations containing a small parameter. *Mathematical Sbornik, 22*, 193–204. (In Russian).

Tikhonov, A. N. (1952). Systems of differential equations containing a small parameter multiplying the derivative. *Mathematical Sbornik, 31*, 575–586. (In Russian).

Van Dyke, M. (1964). *Perturbation Methods in Fluid Dynamics*. Academic Press, New York.

Vasil'eva, A. B. (1963). Asymptotic behaviour of solutions to certain problems involving nonlinear differential equations containing a small parameter multiplying the highest derivative. *Russian Mathematical Surveys, 18*, 13–81.

Vogel-Prandtl, J. (2004). *Ludwig Prandtl*. The International Centre for Theoretical Physics, Trieste, Italy.

Wasow, W. (1965). *Asymptotic Expansions for Ordinary Differential Equations*. Wiley-Interscience, New York.

Williams, H. E. (2008). An asymptotic solution of the governing equation for natural frequencies of a cantilevered, coupled beam model. *Journal of Sound and Vibration, 312*, 354–359.

Young, P. C. (2011). *Recursive Estimation and Time-Series Analysis*. Springer-Verlag, Berlin.

APPENDIX

HISTORICAL REMARKS ON THE DISCIPLINES OF SINGULAR PERTURBATIONS AND AUTOMATIC CONTROL

Singular perturbation methods of differential equations have been the mainstay of modelling dynamical phenomena with slow and fast dynamics ever since the early days of fluid mechanics over a century ago (O'Malley, 2014). Early modelling successes of singular perturbations were in the areas of fluid mechanics, aerodynamics, heat transfer, the skin effect of electrical conductors, and the edge effect in buckling thin elastic plates. The modelling thereby of the so-called 'boundary layer effect' was transformative in these diverse subject areas. Today, wherever slow and fast dynamics are present in a dynamical model, one can be sure that singular perturbation methods will have an indispensable role to play. Indeed, dynamical phenomena possessing both slow and fast modes are to be found almost everywhere.

Automatic control, on the other hand, has had a rather different history. At its heart, automatic control is an engineering design methodology whereby overall control objectives are to be met in the face of dynamical constraints on the control system model. Nowhere is this better exemplified than in the highly successful frequency response methods for control (MacFarlane, 1979). The control design model may be crude, but the frequency response approach allows for this design model uncertainty in meeting control objectives. Moreover, Nyquist-Bode methods cater for transients at any frequency, low or high, in their polar and graphical plots.

So, the disciplines of singular perturbation methods and automatic control were quite different. Singular perturbation methods centred on modelling in the time domain, whereas automatic control centred on engineering design in the frequency domain. This was about to change with the space exploration programmes of the 1950s and 1960s (MacFarlane, 1979).

These prestigious national space programmes demanded precision; precision in modelling, precision in measurement, and precision in guidance, in what was a windless environment. The new control objectives would be guidance and trajectory optimisation on the basis of more accurate time-domain differential equation models. This new differential-equation medium was propitious for the introduction of singular perturbation techniques to the automatic control community that had hitherto been the province of dynamicists (Bagirov, Vasiľeva and Imanaliev, 1967; Kokotović and Sannuti, 1968). After all, what dynamically interesting system does not possess both slow and fast modes? In this way, a substantial body of work rapidly built up on the modelling, analysis and design for control of singularly perturbed systems almost entirely in the time domain (Kokotović and Khalil, 1986; Kokotović, Khalil and O'Reilly, 1999).

With the subsidence of space exploration programmes in the 1970s and 1980s, however, there was a re-emergence of interest in the frequency-domain approach to automatic control (MacFarlane, 1979). This suggested that a singularly perturbed linear time-invariant system, exhibited in transfer-function form, would possess both low-frequency and high-frequency modes (Porter and Shenton, 1975). Indeed, had one looked, small stray capacitance in high-frequency amplifiers (Friis and Jensen, 1924) and electrical networks (Bode, 1945) had already been identified as an oscillation stability issue from a frequency-domain standpoint as early as the 1920s. Other frequency-domain work followed, but it was the important paper of Luse and Khalil (1985) that first approached the two-frequency-scale problem directly, and more generally, in the frequency domain. There, it was defined that for a system to be two frequency scale, the system transfer-function matrix should be proper and the system poles should cloister into two disjoint sets in separate frequency scales. Kokotović, Khalil and O'Reilly (1999) are also notable for developing for the first time, in a more physical way, a more accurate corrected low-frequency model, albeit from a state-space standpoint.

Other extensions of the two-frequency-scale work of Luse and Khalil (1985) followed. Luse (1986, 1988) extends it to multiple time scales. Oloomi and Sawan (1995) investigate zeros, while Oloomi and Shafai (2004) provide the corresponding impulse response for a stable two-frequency transfer function. More recently, the work of Shaker (2009) deals with a frequency-domain generalised singular perturbation method where the importance of accuracy of approximation is stressed, paying particular regard to the preservation of system phase information. Other application works of note are Chaplais and Alaoui (1996), Hu and Wang (2015) and Cavalcanti, Kienitz and Kadirkamanathan (2018) which consider a transfer-function system model in cascade form that is scaled in a manner similar to that developed in Chapter 3 and Chapter 4.

While the foregoing discussion of two-frequency-scale systems primarily addresses itself to conventional automatic control where the focus is naturally on corrected models in the low-frequency scale, other situations exist where the high-frequency scale is of particular interest. For instance, the controls of power electronic devices operate in the high-frequency electromagnetic band, typically anything from a few kilohertz to a few megahertz. On the other hand, the power system within which these power electronic devices are embedded operates in the low-frequency electromechanical band, typically 1 Hz. In other words, power electronically controlled electrical transmission systems are intrinsically two-frequency-scale systems with a very wide separation of frequency scales. Take for example the control of a high-voltage DC (HVDC) converter connected to a weak alternating current (AC) network (O'Reilly, Wood, and Osauskas, 2003). Here the DC current controllers operate in the high-frequency electromagnetic band whereas the rest of the power system is modelled as a quasi-steady-state resonant RLC network.

Historically, this two-frequency-scale situation is very similar to two weakly coupled oscillators (Appleton, 1923) where the weaker oscillator can be represented by such a resonant RLC network; see also Francke, Pogromsky and Nijmeijer (2020) for recent interest in the phenomenon. Although weakly coupled in the high-frequency scale, the two oscillators synchronise in oscillation frequency in the low-frequency scale. More generally, for larger interconnected systems such as power systems, the two-frequency-scale behaviour is one that arises from weakly coupled interconnected dynamical systems. There, the overall interconnected system dynamics evolve together in the low-frequency scale while the local individual system dynamics have dissipated in the high-frequency scale (Peponides, Kokotović and Chow, 1982; Kokotović and Khalil, 1986). Dynamically, the weak interconnections become strong in the low-frequency scale, and the system as a whole synchronises.

INDEX